ADVANCES AND TRENDS IN GEODESY, CARTOGRAPHY AND GEOINFORMATICS

T0187591

PROCEEDINGS OF THE X[TH] INTERNATIONAL SCIENTIFIC AND PROFESSIONAL CONFERENCE ON GEODESY, CARTOGRAPHY AND GEOINFORMATICS, DEMÄNOVSKÁ DOLINA, LOW TATRAS, SLOVAKIA, 10–13 OCTOBER 2017

Advances and Trends in Geodesy, Cartography and Geoinformatics

Editors

Soňa Molčíková, Viera Hurčíková, Vladislava Zeliznaková & Peter Blišťan

Institute of Geodesy, Cartography and Geographical Information Systems, Technical University of Košice, Slovakia

CRC Press
Taylor & Francis Group
Boca Raton London New York

CRC Press is an imprint of the
Taylor & Francis Group, an **informa** business

A BALKEMA BOOK

CRC Press
Taylor & Francis Group
6000 Broken Sound Parkway NW, Suite 300
Boca Raton, FL 33487-2742

First issued in paperback 2020

© 2018 by Taylor & Francis Group, LLC
CRC Press is an imprint of Taylor & Francis Group, an Informa business

Typeset by V Publishing Solutions Pvt Ltd., Chennai, India

ISBN-13: 978-1-138-58489-1 (hbk)
ISBN-13: 978-0-367-73486-2 (pbk)

Visit the Taylor & Francis Web site at
http://www.taylorandfrancis.com

and the CRC Press Web site at
http://www.crcpress.com

Advances and Trends in Geodesy, Cartography and Geoinformatics – Molčíková et al. (Eds)
© 2018 Taylor & Francis Group, London, ISBN 978-1-138-58489-1

Table of contents

Advances and Trends in Geodesy, Cartography and Geoinformatics – Molčíková et al. (Eds)
© 2018 Taylor & Francis Group, London, ISBN 978-1-138-58489-1

Preface

The anniversary 10th International Scientific and Professional Conference Geodesy, Cartography and Geoinformatics (GCG 2017) was organized under the auspices of the Faculty of Mining, Ecology, Process Control and Geotechnologies, Technical University of Košice, Slovakia. The co-organizers of this conference were the Faculty of Science of the Pavol Jozef Šafárik University in Košice (Slovakia), the Faculty of Civil Engineering of Slovak University of Technology in Bratislava (Slovakia), the Faculty of Civil Engineering of Czech Technical University in Prague (Czech Republic), the Polish Polytechnic of Świętokrzyskie in Kielce (Poland), AGH University of Science and Technology in Krakow (Poland), Upper Nitra Mines Prievidza, plc. (Slovakia) and the Slovakian Mining Society (Slovakia). The conference was held October 10–13, 2017 in the Low Tatras, Slovakia.

The purpose of the conference was to facilitate a meeting that would present novel and fundamental advances in the field of geodesy, cartography, and geoinformatics for scientists, researchers, and professionals around Middle Europe. Conference participants had the opportunity to exchange and share their experiences, research, and results with other colleagues, experts, and professionals. Participants also had the possibility to extend their international contacts and relationships for furthering their activities.

The conference focused on a wide spectrum of topics and areas of geodesy, cartography, and geoinformatics as listed below:

1. Surveying and mine surveying
   - Geodetic networks, data processing
   - Engineering surveying and deformation measurement
   - Photogrammetry and remote sensing
   - Real estate cadastre and land consolidation

2. Geodetic control and geodynamics
   - Cosmic and satellite geodesy, theory and applications
   - Vertical reference systems
   - Absolute and relative gravimetry

3. Cartography and Geoinformatics
   - Collection and processing of spatial data,
   - Data standards, infrastructure, metadata, geodatabases
   - Spatial analyses and modeling
   - Development and application of methods and models for spatial processes
   - Digital cartographic systems
   - 3D visualization of spatial data, publishing data on the internet, virtual reality

32 papers from countries V4 were accepted for publication in the conference proceedings of the Conference. Each of the accepted papers was reviewed by selected reviewers in accordance with the scientific area and orientation of the papers.

The editors would like to express their many thanks to the members of the Organizing and Scientific Committees. The editors would also like to express a special thanks to all reviewers, sponsors and conference participants for their intensive cooperation to make this conference successful.

Soňa Molčíková

Viera Hurčíková

Vladislava Zelizňaková

Peter Blišťan

Advances and Trends in Geodesy, Cartography and Geoinformatics – Molčíková et al. (Eds)
© 2018 Taylor & Francis Group, London, ISBN 978-1-138-58489-1

Committees of GCG 2017

SCIENTIFIC COMMITTEE

Prof. Jaroslav Hofierka, *Pavol Jozef Šafárik University in Košice, Slovakia*
Prof. Alojz Kopáčik, *Slovak University of Technology in Bratislava, Slovakia*
Prof. Janka Sabová, *Technical University of Košice, Slovakia*
Prof. Jacek Szewczyk, *Polish Polytechnic of Świętokrzyskie in Kielce, Poland*
Prof. Martin Štroner, *Czech Technical University in Prague, Czech Republic*
Prof. Ján Tuček, *Technical University in Zvolen, Slovakia*
Assoc. Prof. Peter Blišťan, *Technical University of Košice, Slovakia*
Assoc. Prof. Václav Čada, *University of West Bohemia, Plzeň, Czech Republic*
Assoc. Prof. Jana Ižvoltová, *University of Žilina, Slovakia*
Assoc. Prof. Juraj Janák, *Slovak University of Technology in Bratislava, Slovakia*
Assoc. Prof. Branislav Kršák, *Technical University of Košice, Slovakia*
Assoc. Prof. Katarína Pukanská, *Technical University of Košice, Slovakia*
Assoc. Prof. Rudolf Urban, *Czech Technical University in Prague, Czech Republic*
Dr. Piotr Parzych, *AGH University of Science and Technology in Krakow, Poland*
Dr. Branislav Droščák, *Geodetic and Cartographic Institute Bratislava, Slovakia*
Katarína Leitmannová, *Geodesy, Cartography and Cadastre Authority of Slovak Republic, Slovakia*

ORGANIZING COMMITTEE

Dr. Soňa Molčíková
Dr. Viera Hurčíková
Dr. Vladislava Zelizňaková

LIST OF REVIEWERS

M. Bajtala, K. Bartoš, J. Beck, M. Bindzár, M. Bindzárová Gergeľová, P. Blišťan, D. Bobíková,
J. Derco, R. Ďuračiová, J. Erdélyi, R. Fencík, M. Fraštia, B. Hábel, V. Hurčíková, J. Ižvoltová,
K. Janečka, J. Kaňuk, Ľ. Kovanič, Ž. Kuzevičová, P. Kyrinovič, K. Kyšeľa, S. Labant,
M. Marčiš, S. Molčíková, J. Papčo, P. Parzych, K. Pukanská, J. Pukanský, Š. Rákay, P. Repáň,
J. Sabová, M. Stričík, M. Štroner, J. Szewczyk, M. Talich, R. Urban, V. Zelizňaková

Part A: Surveying and mine surveying

Impact of a horizontal refraction on a geodetic network adjustment

Marek Bajtala & Štefan Sokol
Slovak University of Technology in Bratislava, Bratislava, Slovakia

ABSTRACT: The normally used procedure for the geodetic control network adjustment is the least squares method. Assuming that random measurement errors come from the normal probability distribution, estimates obtained using the least squares method are optimal (unbiased and efficient estimate). However, in real conditions, outliers may be part of a set of the measurements, such as the impact of a horizontal refraction. These outliers are causing a violation of the assumption of normality and, therefore, the estimates derived using least squares method lose their optimal properties. For high-quality processing of accurate geodetic control networks, it is necessary to detect outliers and reduce their influence on the resultant estimate using a suitable concept. In the submitted paper, a horizontal geodetic network based on the terrestrial measurements is processed using the least squares method with refraction model and the robust statistical methods. Chosen model of the estimation of horizontal refraction is based on the theoretical basis of defining refraction between two points and the occurrence of refractive blocks in a horizontal geodetic network. The results of experiments indicate that the most accurate results, thus the lowest incidence of the horizontal refraction, are permanent during cloudy weather. If the measurement is done during sunny weather, it is more appropriate to use the least squares method supplemented with a horizontal refraction model to process a horizontal geodetic network.

1 INTRODUCTION

For processing a control network, it is mostly assumed that measured values are affected by the errors, which have a character of normal distribution of the probability $N(\mu, \sigma^2)$. However, measured quantities come from a strictly normal distribution only if gross errors, mistakes and systematic errors are filtered out (Gašincová 2011b). Random character of surveying measurements is conditioned by the measurement itself, where multiple measurements of a particular quantity deliver similar, however different results. Different results may also be caused by the impact of refraction, respective to the impact of current weather conditions during the measurement.

This paper deals with the processing of three experimental measurements in the horizontal control network using a model that takes into account the impact of horizontal refraction and its results are linked to the results obtained in the publications (Bajtala, 2006), (Bajtala, 2007).

2 LINEAR MODEL CONSIDERING REFRACTION

Creation of linear model considering refraction (LMCR) is based on a standard procedure using the least square method (LSM) which is expanded by the effect of horizontal refraction (Bajtala, 2006). The given model can be written as follows:

$$\mathbf{Y} = \mathbf{f}_0 + (\mathbf{F}, \mathbf{G}, \mathbf{H}) \begin{pmatrix} \delta\boldsymbol{\beta} \\ \delta\boldsymbol{\alpha} \\ \mathbf{r} \end{pmatrix} + \boldsymbol{\varepsilon}; \quad \mathrm{Var}(\mathbf{Y}) = \boldsymbol{\Sigma}, \tag{1}$$

where estimate of the unknown parameters $(\delta\hat{\boldsymbol{\beta}}^{\mathrm{T}}, \delta\hat{\boldsymbol{\alpha}}^{\mathrm{T}}, \hat{\mathbf{r}}^{\mathrm{T}})^{\mathrm{T}}$ can be calculated:

$$
\begin{pmatrix} \delta\hat{\boldsymbol{\beta}} \\ \delta\hat{\boldsymbol{\alpha}} \\ \hat{\mathbf{r}} \end{pmatrix} = \left[\begin{pmatrix} \mathbf{F}^{\mathrm{T}} \\ \mathbf{G}^{\mathrm{T}} \\ \mathbf{H}^{\mathrm{T}} \end{pmatrix} \sum{}^{-1}(\mathbf{F},\mathbf{G},\mathbf{H}) \right]^{-1} \begin{pmatrix} \mathbf{F}^{\mathrm{T}} \\ \mathbf{G}^{\mathrm{T}} \\ \mathbf{H}^{\mathrm{T}} \end{pmatrix} \sum{}^{-1}(\mathbf{Y}-\mathbf{f}_0),
\tag{2}
$$

where $\mathbf{f}_0 = \mathbf{f}(\beta_0,\, \boldsymbol{\alpha}_0,\, r_0)$, $\mathbf{F} = \partial\mathbf{f}(\boldsymbol{\beta})/\partial\boldsymbol{\beta}^{\mathrm{T}}|_{\beta=\beta_0}$, $\mathbf{G} = \partial\mathbf{f}(\boldsymbol{\alpha})/\partial\boldsymbol{\alpha}^{\mathrm{T}}|_{\alpha=\alpha_0}$ and $\mathbf{H} = \partial\mathbf{f}(\mathbf{r})/\partial\mathbf{r}^{\mathrm{T}}|_{r=r_0}$.
Final covariance matrix of the unknown parameters has the following form:

$$
Var\begin{pmatrix} \delta\hat{\boldsymbol{\beta}} \\ \delta\hat{\boldsymbol{\alpha}} \\ \hat{\mathbf{r}} \end{pmatrix} = \left[\begin{pmatrix} \mathbf{F}^{\mathrm{T}} \\ \mathbf{G}^{\mathrm{T}} \\ \mathbf{H}^{\mathrm{T}} \end{pmatrix} \sum{}^{-1}(\mathbf{F},\mathbf{G},\mathbf{H}) \right]^{-1}.
\tag{3}
$$

By means of the above described model, we can estimate individual horizontal refractions. Analyses of the identifiability of the estimated horizontal refraction can be carried out by testing null hypothesis. Analysis is based on the knowledge of the sub-matrix of the refraction components $[\mathbf{H}^{\mathrm{T}}\boldsymbol{\Sigma}^{-1}\mathbf{H}] = var(\hat{\mathbf{r}})$ (Bajtala, 2006). The null hypothesis for rejecting the existence of horizontal refraction can be written as follows (Kubáčková, 2000):

$$
\hat{\mathbf{r}}^{\mathrm{T}}[(33)]^{-1}\hat{\mathbf{r}} \overset{H_0}{\sim} \chi_{\mathrm{p}}^2(\mathbf{0}),
\tag{4}
$$

where $\chi_{\mathrm{p}}^2(\mathbf{0})$ is a random variable with chi-squared distribution and p degrees of freedom, where p is the number of considered refraction coefficients in the control network.

3   ROBUST STATISTICAL METHODS

Robust estimation can be applied in basic or extended linear models. M-estimates are mostly used in surveying because they are flexible also for generalized linear models (Hampacher, 2011), (Třasák, 2011). Robust M-estimates, implemented in LSM, are based on a consecutive iterative calculation. To create iterative calculation, which more effectively searches and detects outliers i.e. defines robust weight matrix W for each iteration, it is necessary to calculate standardized correction $\hat{v}_i$ from the correction estimated by LSM. Standardization in LSM is adequate to ration of the calculated correction v_i and particular standard error σ_i, because real errors are unknown and their estimates are corrections v_i:

$$
\hat{v}_i = \frac{v}{\sigma_i}.
\tag{5}
$$

Consequently, robust matrix $\mathbf{W}$ can be defined as follows:

$$
\mathbf{W} = diag(w_1, w_2, \dots w_n).
\tag{6}
$$

Calculation is iterative, as null iteration is considered an LSM estimate that is not robust. Estimated value of the correction v_i using LSM is considered as equally precise values against other corrections, i.e. the correction would have theoretical value $w_i(v_i) = 1$. In such a case, that shape of influence function $\psi(v_i)$ is not bounded, i.e. it is not robust because does not take into account any outliers, respectively intervals of values. Then, it can be stated that corrections are equal and equally effect the calculation of the accuracy and LSM estimate. Consequently, for 1. iteration enters into the normal equations the robust weight matrix W, which acquired certain values depending on their equation and detected outliers, and M-estimate of the first

4

estimation is created. By means of standardized corrections $\hat{v}_i^n$ and consequent iterations are calculated new robust weights w_i^{n+1}, which form weight matrix $\mathbf{W}^{n+1}$ a consequently also M-estimate:

$$\Delta\hat{\mathbf{\Theta}}^{n+1} = \left(\mathbf{A}^T \sum{}^{-1} \mathbf{W}^{n+1} \mathbf{A} \right)^{-1} \mathbf{A}^T \sum{}^{-1} \mathbf{W}^{n+1}, \tag{7}$$

where $\mathbf{A}$ is a matrix of the above described model of refraction and is formed by sub-matrixes $\mathbf{F}$, $\mathbf{G}$ and matrix $\mathbf{H}$. Of existing M-estimates has been used for calculation—estimate of Cauchy's distribution (Hampacher, 2011), (Třasák, 2011), (Gašincová, 2011a), (Giu 1998) and we will continue to designate this model as LMCR-Cauchy.

4 EXPERIMENTAL MEASUREMENTS

All three experimental measurements were carried out on the channel of the Gabčíkovo water work, between Vojka nad Dunajom and Kyselica harbor. Due to different surfaces in this locality (asphalt, water, and grass on the slopes agricultural land) is an assumption of the presence of the horizontal refraction.

Atmospheric conditions during the individual experimental measurements are found in Table 1.

The horizontal control network comprised 9 points on both sides of the channel (Figure 1). All points were stabilized with pillars. Measurement was carried out by Leica

Table 1. Atmospheric condition during experimental measurements.

Experiment	Temperature [°C]	Pressure [hPa]	Humidity [%]	Wind speed and direction [km/h]	Cloudiness
1.	7–8	1003,6	79–71	30–60, North	High clouds
2.	14–17	953,5	57–58	30, South	Low clouds
3.	18–22	999,9	50–60	No wind	Sunny

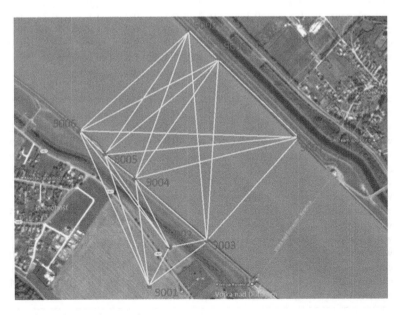

Figure 1. Horizontal control network.

TS30 and a set of Lieca prisms. Distances and directions were measured in 6 sets using automation target recognition function (ATR). The result was multiple measurement of 60 horizontal directions and 30 distances, which were used for calculation.

Calculation was performed by means of the above described linear model considered refraction LMCR-Cauchy. The objective of the experimental measurement was verification of a suitability of defined LMCR-Cauchy model in a horizontal control network and estimation of the impact of horizontal refraction on a particular direction, respective to estimation of horizontal refraction and their characteristic of accuracy. Variance components of the instrument were estimated as well, consequently, analysis of the identifiability of the existence of refraction was realized.

Results of the estimated horizontal refraction with the related characteristic of accuracy, which were calculated by means of LMCR-Cauchy model, are given in Table 2. The table also includes calculated coefficients of identifiability of the existence of refraction, assessed by the coefficient of identifiability of the existence of refraction, calculated according to equation (4).

Table 2. Estimated horizontal refraction and their standard deviations with coefficient of identifiability.

Sight	Surface	Refraction [cc]			Standard deviation [cc]			Coefficient of identifiability $k_r^* = 3{,}84$		
		E-1	E-2	E-3	E-1	E-2	E-3	E-1	E-2	E-3
1–2	Agricultural	−1,7	−1,4	0,8	0,9	0,8	2,0	3,1	2,9	0,0
1–3	land-slope	−1,6	−0,6	0,6	1,1	1,0	2,0	2,0	0,1	0,0
1–4	land-slope	−1,5	−1,4	−0,7	1,2	1,0	2,0	1,1	2,8	0,9
1–5	land-slope	−1,1	0,0	**−4,4**	1,2	1,1	2,1	1,0	0,1	5,6
1–6	land-slope	−0,2	0,5	1,8	1,3	1,1	2,1	0,0	0,0	0,3
2–3	slope	**3,0**	−0,5	0,7	0,8	0,7	1,9	12,5	0,1	0,4
2–4	slope	−0,5	−0,3	−1,2	0,9	0,8	1,9	0,1	0,6	0,6
2–5	slope	0,1	−0,5	1,1	1,0	0,8	2,0	0,0	1,2	0,6
2–6	slope	1,4	0,1	1,0	1,0	0,9	2,0	2,3	0,1	0,5
3–4	embankment	0,6	0,3	**16,7**	0,8	0,7	1,8	0,7	0,2	77,3
3–5	embankment	1,0	1,2	**12,6**	0,9	0,7	1,8	0,7	0,5	46,4
3–6	embankment	**2,2**	0,6	**8,3**	0,9	0,8	1,9	5,2	0,0	19,3
3–7	water surface	0,0	−1,7	−0,2	1,1	1,0	1,9	0,0	6,4	0,2
3–8	water surface	−1,8	−0,6	0,3	1,0	0,9	1,9	5,1	1,7	0,0
3–9	water surface	**−0,7**	**−2,7**	1,4	1,1	1,0	1,9	0,0	5,1	0,1
4–5	embankment	−0,2	0,6	1,7	0,8	0,6	1,8	0,5	0,6	1,7
4–6	embankment	0,4	−0,8	**5,7**	0,8	0,7	1,8	0,1	1,8	12,0
4–7	water surface	2,0	−0,4	1,2	1,1	0,9	1,9	3,8	0,2	0,4
4–8	water surface	0,7	−0,7	0,6	0,9	0,8	1,9	0,0	0,5	0,2
4–9	water surface	0,1	−1,3	1,6	1,0	0,9	1,9	0,7	0,0	0,7
5–6	embankment	0,6	0,0	**4,7**	0,8	0,6	1,8	0,9	0,0	7,0
5–7	water surface	0,8	−1,8	0,2	1,1	1,0	1,9	1,3	3,7	0,1
5–8	water surface	0,9	−0,5	1,9	0,9	0,8	1,9	0,5	0,1	0,6
5–9	water surface	−1,1	−0,7	3,0	1,0	0,9	1,9	0,0	0,4	1,6
6–7	water surface	−1,4	−0,5	0,8	1,1	1,0	2,0	1,5	0,3	0,0
6–8	water surface	−0,6	−0,5	−0,4	1,0	0,8	1,9	1,2	0,1	0,2
6–9	water surface	−0,4	0,1	1,3	1,1	0,9	2,0	0,2	2,0	0,1
7–8	embankment	**2,8**	−0,4	**−18,3**	1,0	0,9	1,9	5,1	0,1	89,6
7–9	embankment	**4,0**	0,0	**−23,8**	1,1	1,0	1,9	19,5	1,4	149,2
8–9	embankment	−0,3	0,5	**−5,3**	0,9	0,8	1,9	1,2	3,9	8,4

As can be seen from Table 2, in the first experimental measurement in high cloudiness and with a relatively strong north wind, horizontal refraction was detected at four sights ($2,2^{cc}$ to $4,0^{cc}$). For other sights, the estimated refraction angle values were within the precision of refractive angle determination (around 1^{cc}). Similar results can be seen in the second experimental measurement in low cloud and mild southern wind, where even horizontal reflection was detected on only one sight ($-2,7^{cc}$). Based on the results of the first and second experimental measurements, we can state that the horizontal refraction was minimal. In the third experimental measurement in clear and sunny weather, the horizontal refraction was detected on nine sights. However, the results of detected refraction angles significantly exceeded the precision of the refraction angle estimate. Further, we can see that the sights on which the refraction was detected were mainly over the barrier. This also confirms a theoretical assumption about the presence of the horizontal refraction in the refraction blocks published in (Bajtala, 2006), where is stated the assumption that impact of horizontal refraction in the vicinity of water streams will affect sight parallel with the embankment of the water stream. Refraction was not identified on the sight above the water surface.

For the third experimental measurement, in which the effect of the horizontal refraction is best absorbed, the following Table 3 shows the estimated coordinates and their standard deviations as calculated in the LSM model and in the LMCR-Cauchy model. In the last column, are coordinate differences that are calculated as the differences between estimate coordinates of the LSM model and the LMCR-Cauchy model.

It can be seen from the differences between classical LSM and LMCR-Cauchy model the differences are up to 1 mm. The accuracy of the estimation of the coordinates is also in the two models of up to 1 mm. In the first and second experimental measurement, the coordinate differences varied around 0,5 mm, which is logical with respect to the minimal effect of refraction.

The estimation of the variance components (only direction) of instruments in the third experiment was 6^{cc} in the LSM model and 2^{cc} in the LMCR-Cauchy model. The 2^{cc} value of the LMCR-Cauchy model also corresponds to the classic calculation of standard deviation of the measured directions in 6 groups ($1,85^{cc}$).

Table 3.   Comparison of coordinate differences between LSM and LMCR-Cauchy in experimental measurement E3.

Experimental measurement E3					
	LSM		LMCR-Cauchy		
Point	X, Y [m]	σ [mm]	X, Y [m]	σ [mm]	ΔX, ΔY [mm]
X9001	5040,2862	0,9	5040,2852	0,7	**−1,0**
Y9001	712,0377	0,7	712,0372	0,7	**−0,5**
X9002	5080,6816	0,7	5080,6806	0,4	**−1,0**
Y9002	882,2929	0,7	882,2924	0,5	**−0,5**
X9004	5518,9368	0,6	5518,9358	0,5	**−1,0**
Y9004	999,3316	0,6	999,3313	0,4	**−0,3**
X9005	5619,5743	0,7	5619,5734	0,5	**−0,9**
Y9005	1000,1269	0,5	1000,1266	0,3	**−0,3**
X9006	5718,8546	0,7	5718,8538	0,5	**−0,8**
X9007	4994,9794	1,1	4994,9783	1,0	**−1,1**
Y9007	1556,1722	1,0	1556,1717	1,0	**−0,5**
X9008	5493,1234	1,1	5493,1229	1,0	**−0,5**
Y9008	1598,7979	0,7	1598,7980	0,5	**0,1**
X9009	5692,3870	1,2	5692,3866	1,1	**−0,4**
Y9010	1616,2313	0,8	1616,2308	0,7	**−0,5**

5 CONCLUSION

These three experimental measurements point to the fact that at present, if we want to achieve the most accurate measurement results in the geodetic network, the limitation factor is the environmental impact. The first and second experimental measurements confirmed that when geodetic measurements are carried out under suitable conditions (constant temperature, mild wind, cloudy weather), the measurement is influenced by climatic influences (by refraction) minimally. In this case, the LSM model is sufficient to calculate. However, if the measurement is carried out during sunny weather, where different air layers are heated by varying temperature gradients, the LSM model may not be sufficient to correctly calculate the parameters of the geodetic network. An example of this is the third experimental measurement where the horizontal refraction values exceeded 20^{cc}. By introducing the LMCR-Cauchy model, it is possible to detect refraction and thereby eliminate its impact.

ACKNOWLEDGEMENTS

This paper is part of grant no. 037STU-4/2016 "Modernization and Development of Technological Skills in teaching Surveying and Photogrammetric".

REFERENCES

Bajtala, M. Vplyv horizontálnej refrakcie na geodetické merania. Doctoral thesis, Katedra geodézie SvF STU v Bratislave, Slovakia, 2006, p. 86.

Bajtala, M., Sokol, Š., Ježko, J. Odhad parametrov geodetickej siete s uvážením vplyvu horizontálnej refrakcie. In: Acta Montanistica Slovaca Year 12 (2007), special no. 3, pp 301–310. ISSN 1335–1788.

Gašincová, S., Gašinec, J., Weiss, G., Labant, S., Application of robust estimation methods for the analysis of outlier measurements. In: Geo Science Engineering, vol. 57, no. 3, Czech Republic, 2011a, pp 14–29. ISSN 1802–5420.

Gašincová, S., Gašinec, J., Staňková, H., Černota, P., Porovnanie MNŠ alternatívnych odhadovacích metód pri spracovaní výsledkov geodetických mérní, In: Sborník referátů XVIII. Conference SDMG, Praha, Czech Republic, 2011b, pp 40–51, ISBN 978-80-248-2489-5.

Giu, Q., Zhang, J., Robust biased estimation and its applications in geodetic adjustments. In: Journal of Geodesy, USA, 1998, pp 7–8, pp 430–435, ISSN 0949–7714.

Hampacher, M., Štroner, M., Zpracování a analýza meření v inženýrské geodézii. České vysoké učení technické, Fakulta stavební ČVUT v Prahe, Czech Republic, 2011, p. 312, ISBN 978-80-01-04900-6.

Kubáčková, L., Kubáček, L., Statistika a metrologie, Univerzita Palackého v Olomouci, Czech Republic, 2000.

Třasák, P., Štroner, M., Detection of Outliers in the Adjustment of Accurate Geodetic Measurement. In: Ingeo 2011. Zagreb: University of Zagreb, Faculty of Geodesy, Croatia, 2011, vol. 1, pp 105–116, ISBN 978-953-6082-15-5.

Advances and Trends in Geodesy, Cartography and Geoinformatics – Molčíková et al. (Eds)
© *2018 Taylor & Francis Group, London, ISBN 978-1-138-58489-1*

3D digital mapping of cave spaces in Slovakia by terrestrial laser scanning

K. Bartoš, K. Pukanská, Š. Rákay & J. Sabová
Institute of Geodesy, Cartography and GIS, Technical University of Košice, Košice, Slovakia

P. Bella
*State Nature Conservancy of the Slovak Republic, Slovak Caves Administration,
Hodžova, Liptovský Mikuláš, Slovakia*

ABSTRACT: Slovakia belongs to countries with wonderful and peculiar nature. One of the remarkable natural phenomena, which appeal to the general public as well as various scientists by its sights, are caves. Compared to other natural phenomena, they are characterised by many distinctive and unique features, which enhance the mystique of the underground. At present, caves have an educational character and an important role in environmental education; however, they also have a significant position in tourism. Approximately 7100 caves are known in Slovakia, including shorter ones with the character of overhang. The most registered caves are situated in national parks Slovak Karst, Low Tatras, Veľká Fatra, West, High and Belianske Tatras and in the Spiš-Gemer Karst. Nowadays, the speleological mapping in Slovakia does not have any unified rules. Therefore, final maps of individual cave spaces may vary in their details, accuracy, as well as their content. If a speleologist realises mapping, then the base map is usually less accurate, but its content is rich regarding morphology and speleology. However, if a surveyor performs the mapping, then the base map can be relatively accurate but does not always capture all important morphological features. The objective of this paper is to highlight the possibilities of non-contact surveying technologies for mapping cave spaces, to help create unified rules for speleological mapping combining the rich content of morphological data by standard speleological mapping and high accuracy and details by geodetic measurements.

1 INTRODUCTION

A cave can be defined as a natural underground cavity space, which can be formed by dissolving, or weathering of base rock in karst areas, as a gas cavity in volcanic rocks, or as an abrasive cave on coasts. However, the formation of caves represents a summary of complex processes, with climatic conditions and rock subsoil playing a crucial role (Hofierka et al., 2016). Several genetic types of caves are a demonstration of considerable geodiversity of natural values in the area of Slovak Republic. They are a remarkable example of the development of underground karst phenomena (some of them also internationally). The natural value of caves is also related to the character of their filling, especially rare calcite and aragonite sinter forms. Moreover, caves have always attracted a man by its mystique. Some of them were populated or used as sacrificial places or shelters in the past. Several caves in Slovakia are among the most important archaeological sites (Domica Cave, Ardovská Cave, and others) (Bella, 2005).

In Slovakia, there are currently 13 show caves provided by the Slovak Caves Administration—SCA (Belianska Cave, Brestovská Cave, Bystrianska Cave, Demänovská Cave of Liberty, Demänovská Ice Cave, Dobšinská Ice Cave, Domica Cave, Driny Cave, Gombasecká Cave, Harmanecká Cave, Jasovská Cave, Ochtinská Aragonite Cave and Važecká Cave); 5 other show caves operated by other subjects and 30 caves freely open to the public. Overall, over 7100 caves, including shorter caves with the character of overhang, are known. In Slovakia, the protection of caves is implemented under the Act no. 543/2002 on Nature and Landscape Protection. Caves are protected as natural monuments and most important caves as national natural monuments. Since caves are among the most attacked monuments by humans, they require special protection. Their appropriate and optimal use lies in monitoring the impact of visitors on the natural environment of caves, including its regulation. All show caves are declared as National Natural Monuments. Ochtinska Aragonite Cave, Domica Cave, Gombasecká Cave and Jasovská Cave are included in the World Natural Heritage since 1995. The Dobšinská Ice Cave is inscribed on the World Natural Heritage since November 2000. The protection and operation of show caves in Slovakia are ensured by the SCA in Liptovský Mikuláš, as a professional organisation of nature protection of the Ministry of Environment of the SR (ssj.sk, 2017).

Caves represent specific underground forms of georelief, which attract the attention of various sciences exploring the landscape from different aspects, but mainly speleologists. One of the most frequent activities in speleology is the use of a map. The cave spaces are usually very complex, considerably more complex than surface shapes and objects. Therefore, other common information techniques, like a sketch, are usually not sufficient for speleological activities; it is necessary to use or create a scaled image of cave spaces—the map. The creation of cave maps, or mapping of underground spaces belongs to difficult activities that have a direct impact on practically all speleological activities (Hromas, Weigel, 1988; Hochmut, 1995).

This paper aims to highlight the possibilities of using a non-contact surveying technology—terrestrial laser scanning, in mapping caves in Slovakia and thus help to establish uniform rules for mapping caves, combining the rich content of morphological data of classical speleological mapping and high accuracy of geodetic surveying. The results of these measurements can subsequently be used by speleologists, geologists, geographers, as well as in the area of tourism and other disciplines. Although the use of non-contact surveying technologies like TLS (or digital photogrammetry) for mapping underground spaces is quite new in Slovakia, the use of such technologies in underground spaces is already applied relatively frequently in the world; whether it is an overall or just a partial mapping of cave spaces (Tsakiri et al., 2007; Lerma et al., 2009; Canevese et al., 2011; Gašinec et al., 2012; Amparo Núñez et al., 2013; Silvestre et al., 2013). Since 2011, we have successfully used terrestrial laser scanning for mapping the following caves in Slovakia (Fig. 1):

Figure 1.   The location of caves mapped using TLS in Slovakia since 2011.

Figure 2.   TLS of the Dobšinská Ice Cave and the final point cloud.

10

- Dobšinská Ice Cave (show cave provided by SCA) – 2011 and 2012,
- Medvedia Cave (not accessible to the public) – 2013,
- Dúpnica Cave (not accessible to the public) – 2013,
- Ochtinská Aragonite Cave (show cave provided by SCA) – 2014,
- Belianska Cave (show cave provided by SCA) – 2014.

Due to the inaccessibility of the Medvedia Cave to the public, its complexity and the extent of geodetic activities, the separate paper deals with the survey of the Medvedia Cave.

All laser scanner works were realised using the Leica ScanStation C10 laser scanner (Fig. 2). It is a pulse laser scanner with dual-axis compensator with the precision of modelled surface 2 mm and accuracy of single measurement 6 mm in position and 4 mm in the distance. The range of the scanner is 134 m for 18% albedo of the scanned surface, and the speed of scanning is 50 000 points/sec. Primary processing of point clouds was performed in Leica Cyclone software and subsequent editing and modelling in CloudCompare and Trimble RealWorks software.

2 DOBŠINSKÁ ICE CAVE

The Dobšinská Ice Cave is situated on the south-west edge of the National Park Slovak Paradise in the Spiš-Gemer karst (Fig. 1). The entrance to the cave is on the northern slope of the Duča Hill at an altitude of 969 m a.s.l., 130 m above the bottom of the Hnilec Valley. The cave belongs to the most important glaciated caves in the world, and it is included in the World Heritage UNESCO since 2000. At the same time, it is one of the first electrically illuminated caves in the world—since 1887. However, the Dobšinská Ice Cave is well-known mainly due to an exceptional range of ice filling, which was formed in underground spaces at an altitude of only 920 to 950 m a.s.l. (other important ice caves are mostly located at higher altitudes). The ice filling occurs in the form of floor ice, icefalls, ice stalagmites and pillars. The glaciated area is 9772 m^2 with the ice volume of more than 110 100 m^3. The ice filling is not static; it changes depending on climatic conditions and its gravitational deformations. The main objective of the laser scanning was the survey and subsequent digital modelling of changes in ice filling, not only from a geoscientific and environmental point of view but also in terms of safety and maintenance of visitors' walkways (Bella, 2006; Gašinec et al., 2012).

The laser scanning was realised during the winter period since at that time the loss of ice filling is the most significant due to sublimation and at the same time, it is not yet affected by the spring snow melting. The laser scanning was carried out with the point density 2 × 2 cm at the distance of 30 m. The use of the integrated digital camera to capture the photo-texture was not necessary. Therefore, the final point cloud only contains information on the intensity of the reflected signal. The point cloud with approx. 64 million points was the result of laser scanning. Subsequently, the point cloud was used to generate a MESH model of the ice filling surface. In the next phase, the final MESH model of glaciated parts was compared to models obtained by other methods—tacheometry and photogrammetry; to determine the accuracy of laser scanner when scanning ice surfaces, since there was the expectation that the laser beam will not be reflected from the ice surface, but it should penetrate into the ice (Gašinec et al., 2012).

3 DÚPNICA CAVE

Dúpnica Cave is located on the southwestern edge of the Western Tatras in Slovakia, at the boundary of the Western Tatras and the Chočské vrchy (Fig. 1). The entrance to the cave, measuring approximately 10 m × 1,5 m, is situated at an altitude of 765 m a.s.l. Currently, the total length of the cave is 150 m, while a large part of the cave is formed by a large dome measuring approximately 40 × 35 m and 15 m high. The purpose of laser scanning of the cave dome was a three-dimensional visualisation of the cave, so a more sophisticated and more detailed interpretation of a complex morphology including fault surfaces and lines was possible. This allowed not only to identify fault lines but also to get information on their direction and slope, even in hardly accessible parts of the cave. Detailed data on structural geology

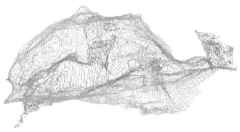

Figure 3. The entrance to the cave and the set of vertical cross-sections through the model.

and morphology were essential regarding completing and refining the existing knowledge of genesis of the cave.

The geodetic measurements of the entrance to the cave, the main cave dome and the section decreasing to the lower and side parts were realised from five scanning stations by terrestrial laser scanner Leica ScanStation C10. The whole scanning was transformed into the national coordinate and vertical system. The given areas were scanned with the density of 2×2 cm, and the final point cloud was subsequently processed into the MESH model (with density reduced to 20×20 cm). The resulting MESH model was graphically illustrated in the form of vertical cross-sections in the direction of two major axes – x and y (Fig. 3).

Based on this 3D digital model of the cave surface of the Dúpnica Cave, it was possible to complete and refine its morphologic characteristics and basic morphometric indicators and determine the spatial distribution of faults. The laser scanning of this cave and its results confirmed that thus obtained a digital 3D model of the cave's surfaces is an important tool for the examination of structural and geological discontinuities, comprehensive and detailed morphology of caves, the creation of cross-sections and derivation of morphometric indicators. Compared with classical speleological mapping, it was possible to determine the orientation of faults even in hardly accessible parts of high walls and ceilings, thus gaining a more illustrative and complex image of structural and geological conditions in the cave (Bella et al., 2015).

4 OCHTINSKÁ ARAGONITE CAVE

The Ochtinská Aragonite Cave is located in Southern Slovakia, on the north-western slope of the Hrádok Hill in the Revúcka Highlands, to the west of Rožňava city (Fig. 1). The cave is featured by various morphologies; it consists of striking linear passage formed along a steep fault and longer irregular labyrinth of passages and halls with many cupolas, ceiling deep or shallow hollows, niches, water level notches and floor looped conduits. The cave is well-known and important due to the richness and variety of aragonite filling, created under specific hydrochemical and climatic conditions in closed underground hollows. The cave was discovered by chance while drilling the geological survey in 1954. Cave development works for an opening to the public started in 1966 by thrilling the access adit 145 m long, which enabled opening the cave to the public in 1972. The total length of the cave is 585 m, while 230 m are accessible.

Since the classic surveying and mapping of bedrock surfaces with irregular and rugged morphologies realised in the past was very lengthy and not very precise, the whole cave was geodetically surveyed by terrestrial laser scanning in June 2014. Geodetic measurement of The Ochtinská Aragonite Cave was realised by TLS from points of the original survey net (Fig. 4). Suitable survey stations were chosen after the terrain reconnaissance, while points of the geodetic control monumented in the concrete sidewalk of the guided route were used as initial surveying points. The spatial coordinates of morphological structures were determined in the national coordinate and vertical system using the terrestrial laser scanner Leica ScanStation C10.

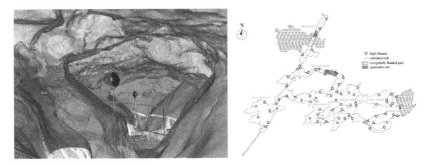

Figure 4.   A part of the final point cloud and the final map of the Ochtinska Aragonite Cave.

Figure 5.   The final point cloud and MESH model of the part of the Belianska Cave.

Overall, more than 121 million points from 52 survey stations monumented in the concrete sidewalk were measured during the laser scanning. The spatial resolution of 2 × 2 cm was achieved. A MESH model with the resolution of points up to 10 cm, longitudinal sections and vertical cross-sections of individual parts of the cave were created by subsequent processing. The result is a detailed map of the whole cave (Fig. 4).

5   BELIANSKA CAVE

The Belianska Cave is situated on the northern slope of the Kobylí Hill in the eastern part of the Belianske Tatry (Fig. 1). It is located in the National Nature Reserve Belianske Tatry on the territory of the Tatra National Park. The cave reaches the length of 3 829 m and the depth of 168 m. Morphologically, it consists of two main, north descending branches, which connect in its upper sub-horizontal part and partially also in the lower, also predominantly sub-horizontal part. The verticality of the cave is complemented by several abysses (Famine Abyss – 34 m, Hell abyss –30 m) and chimneys. The whole cave is situated at altitudes from 865 to 1025 m a.s.l. The entrance to the cave was known for a long time. The entrance parts of the cave were already well known by gold-diggers in the first half of the 18th century, as evidenced by the inscriptions of their names on cave walls. The cave is electrically illuminated since 1896. Nowadays, it is used for revitalising speleoclimatic stays (Bella, Pavlarčík, 2002; Bella et al., 2011).

The geodetic survey of the Belianska Cave represented mainly the terrestrial laser scanning in three main parts of the cave—Pipe Dome, Musical Hall and Discoverer's Dome (Fig. 5). TLS was again realised by the Leica ScanStation C10 laser scanner, from 31 scanning stations, while the whole measurement was transformed into the national coordinate and vertical system. The point cloud containing approx. 200 million of points was the result of laser scanning. Subsequently, the point cloud was used to generate MESH models, vertical cross-sections and profiles of individual parts of the cave (Fig. 5).

6 CONCLUSIONS

Mapping caves and obtaining accurate data about their spatial structure is important not only for their documentation, in tourism, propagation, etc., but also to acquire a comprehensive knowledge of the morphology and genesis of these cave spaces. Conventional mapping techniques and methods do not always provide sufficient information. Unlike them, terrestrial laser scanning is a technology that allows a rapid and accurate survey of detailed morphology, with no need for artificial illumination of these underground spaces. Although financial demands and potential sensitivity of the instrument to difficult conditions in the underground can be a slight disadvantage, the accuracy and detail of the resulting cloud, with a suitable selection of scanning stations and scanning resolution, provides an indispensable tool for structural, geological and morphometric analysis of underground spaces.

REFERENCES

Bella, P.: Protection and Utilization of Caves in Slovakia. *Život. Prostr.*, Vol. 39, No. 2, p. 79–82, 2005. ISSN 0044-4863.

Bella, P.: Morphology of ice surface in the Dobšiná Ice Cave. In Zelinka, J. Ed. *Proceedings of the 2nd International Workshop on Ice Caves* (Demänovská Dolina, May 8–12, 2006). Liptovský Mikuláš, pp. 15–23. ISBN 978-80-8064-279-2.

Bella, P., Bosák, P., Pruner, P., Głazek, J., Hercman, H.: The development of the River Biela Valley in relation to the genesis of the Belianska Cave. Geografický časopis/Geographical Journal, Vol. 63, No. 4, 2011. pp. 369–387. ISSN 0016-7193.

Bella, P., Littva, J., Pukanská, K., Gašinec, J., Bartoš, K.: Využitie terestrického laserového skenovania pri skúmaní štruktúrno-tektonických diskontinuít a morfológie jaskýň: príklad jaskyne Dúpnica v Západných Tatrách. *Acta Geologica Slovaca*, 2015, Vol. 7, No. 2. pp. 93–102. ISSN 1338-5674.

Bella, P., Pavlarčík, S.: Morfológia a problémy genézy Belianskej jaskyne. In Bella, P. Ed. Výskum využívanie a ochrana jaskýň. Zborník referátov z 3. vedeckej konferencie (Stará Lesná, 14–16. 11. 2001). Liptovský Mikuláš, Správa slovenských jaskýň, 2002. pp. 22–35. ISBN 80-8064-145-5.

Canevese, E.P., et al.: Laser scanning technology for the hypogean survey: the case of Santa Barbara karst system (Sardinia, Italy). *Acta Carsologica*, Vol. 40, No. 1, 2011. pp. 65–77. ISSN 0583-6050.

Gallay, M., Kaňuk, J., Hochmuth, Z., Meneely, J., Hofierka, J., Sedlák. V.: Large-scale and high-resolution 3-D cave mapping by terrestrial laser scanning: a case study of the Domica Cave, Slovakia. *International Journal of Speleology*, Vol. 44, No. 3, 2015. pp. 277–291. ISSN 1827-806X. doi:10.5038/1827-806X.44.3.6.

Gašinec, J., Bella, P., Gašincová, S., Imreczeová, A.: Elevation model of floor ice surfaces in the Dobšinská Ice Cave, Slovakia/Digitálny výškový model podlahových ľadových povrchov v Dobšinskej ľadovej jaskyni. *Slovenský kras*, Vol. 50, No. 1, 2012. pp. 31–40. ISSN 1335-6410.

Gašinec, J., Gašincová, S., Černota, P., Staňková, H.: Zastosowanie naziemnego skaningu laserowego do monitorowania logu gruntowego w Dobszyńskiej Jaskini Lodowej. *Inżinieria Mineralna*, 2012, Vol. 13, No. 2. pp. 31–42. ISSN 1640-4920.

Hofierka, J., Hochmuth, Z., Kaňuk, J., Gallay, M., Gessert, A.: Mapovanie jaskyne Domica pomocou terestrického laserového skenovania. *Geografický časopis/Geographical Journal*, Vol. 68, No. 1, 2016. pp. 25–38. ISSN 0016-7193.

Hochmut, Z.: *Mapovanie jaskýň*. Slovenská Speleologická Spoločnosť. Liptovský Mikuláš, 1995. ISBN 80-966963-1-9.

Hromas, J., Weigel, J.: *Základy speleologického mapování*. Praha: Česká speleologická spoločnosť 1988. 135p.

Lerma, J.L., et al.: Terrestrial laser scanning and close range photogrammetry for 3D archaeological documentation: the Upper Palaeolithic Cave of Parpalló´ as a case study. *Journal of Archaeological Science*, 2010, Vol. 37, Issue 3. pp. 499–507. ISSN 0305-4403.

Núñez, A.M., Buill, F., Edo, M.: 3D model of the Can Sadurní cave. *Journal of Archaeological Science*, 2013, Vol. 40, Issue 12. pp. 4420–4428. ISSN 0305-4403.

Silvestre, I., et al.: Framework for 3D data modeling and Web visualization of underground caves using open source tools. *Proceedings of the 18th International Conference on 3D Web Technology*, 2013. ACM New York, NY, USA. Pp. 121–128. ISBN: 978-1-4503-2133-4.

Správa slovenských jaskýň. 2017. *Slovak Caves Administration / Správa slovenských jaskýň*. [ONLINE] Available at: http://www.ssj.sk/sk/jaskyne. [Accessed 14 July 2017].

Tsakiri, M., et al.: 3D Laser Scanning for the Documentation of Cave Environments. *11th ACUUS Conference: "Underground Space: Expanding the Frontiers"*, September 10–13 2007, Athens – Greece.

Advances and Trends in Geodesy, Cartography and Geoinformatics – Molčíková et al. (Eds)
© 2018 Taylor & Francis Group, London, ISBN 978-1-138-58489-1

Precision of angular measurement of total stations Trimble M3

J. Braun
Department of Special Geodesy, Faculty of Civil Engineering, CTU Prague, Czech Republic

ABSTRACT: The aim of the paper is to present the results from testing the angular precision of 12 same total stations Trimble M3. The experimental testing was performed according to international standard ISO 17123-3. Test fields were built in the campus of the CTU in Prague. Full test procedure for horizontal directions and zenith angles was performed. The aim was to obtain a real measurement precision under normal atmospheric conditions. The measurements took place between November 2016 and March 2017. The result is a comparison of the experimental standard deviations of horizontal directions and zenith angles with the manufacturer's standard deviation. The instruments are compared to each other and is determined which instrument is the most precise.

1 INTRODUCTION

Precision of total stations is normally provided by the manufacturers for distance and angle measurement. The distance measurement is characterized by a standard deviation $\sigma_D = A$ mm + B ppm that shows the precision of the distance measurement at one face position of the telescope. The A value represents the size of the systematic and random error component that is not distance dependent. The B value is a distance-dependent error component (Rüeger 1990). Angular measurements—horizontal directions and zenith angles are characterized by the standard deviation σ_{AM}, which expresses the precision of the measurement at one point in both faces position of the telescope (ISO 17123-3, 2005).

The precision stated by the manufacturer must be regularly verified in accredited calibration laboratory. The calibration certificate must be documented for measurements in the area of cadaster, mapping and in contracts for geodetic measurements on construction sites. In addition to official calibration measurements, the instruments can be tested with various supplementary methods (Křemen & Koska 2015). Distance meters are usually tested on pillar baselines with known lengths or on laboratory interferometer baselines (Dvořáček 2016). For testing of angles are created special outdoor bases (Nestorovic 2013, Lambrou & Nikolitsas 2015) or special laboratory testing instruments (Bručas et al. 2006, Bručas et al. 2014). Testing of angular precision generally depends on the requirements and processes in ISO 17123-3.

Determining the real measurement precision of the geodetic instrument is especially important to verify that the device is suited to the measurement task. Most often it is a work in construction site and in the measurement of displacements and deformations (Štroner et al. 2014). If more multiple identical instruments are available, the testing is recommended to obtain the most reliable piece (Štroner & Suchá 2008).

At the Department of Special Geodesy at the Faculty of Civil Engineering CTU in Prague are 12 Trimble M3 total stations for education. Instruments are used for common geodetic measurements throughout the academic year. As part of experimental measurements for bachelor thesis (Kůdela 2017) and for experimental testing of geodetic instruments, all instruments were subjected to complete angle measurement test according to ČSN ISO 17123-3. The aim of the experimental measurement was to verify the precision of the instruments and also to compare the instruments to each other and choose the best one for other experimental measurements.

Trimble M3 (Fig. 1) is a modern small total station. It is used in cadastral, mapping and in simpler work in engineering surveying. The instrument is classical mechanical construction with mechanical infinite clamps. The instrument is controlled by the built-in keyboard and the touch screen. The Trimble Access software uses the Windows interface and it is intuitive and simple to operate.

All 12 total stations have the same parameters according to the manufacturer. Standard deviation of the distance measurement is $\sigma_D = 3$ mm $+ 2$ ppm in prism and reflectorless mode and standard deviation of angle measurement is $\sigma_{AM} = 5''$ (1.5 mgon). The manufacturer also delivers the instruments with angular accuracy $\sigma_{AM} = 1''$, $2''$, $3''$.

The instruments are used in a common measuring environment and regularly serviced. The instruments come from two production series (Table 1). Instruments No. 1–6 are used since 2012 and the last service check was (2/2017). Instruments No. 7–12 are used since 2015.

Figure 1.   Total station Trimble M3.

Table 1.   Serial numbers—Trimble M3.

Total station	Serial number
No. 1	C652352
No. 2	C652355
No. 3	C652356
No. 4	C652359
No. 5	C652360
No. 6	C652369
No. 7	D047349
No. 8	D047353
No. 9	D047379
No. 10	D047399
No. 11	D047410
No. 12	D047418

3 TEST FIELDS

3.1 *Test field for horizontal directions*

Test field was built to meet the requirements of ISO 17123-3 at the CTU campus in Prague. For horizontal directions 5 targets were selected at distances from 150 m to 230 m (Fig. 2). International standard requires distances from 100 m to 250 m. Points were chosen to evenly cover the entire circle. The 4 observed points were signaled by reflective foils and one point was the chimney's corner (Fig. 3). The measurement standpoint was marked by the measuring nail in the pavement.

3.2 *Test field for vertical angles*

For testing the zenith angles the highest building in the area was used—building A of the Faculty of Civil Engineering CTU in Prague (Fig. 4). There were 4 points marked by reflective foils (Fig. 5). Points were chosen to cover as much height and were also angularly spaced apart 30 gon. The standpoint was stabilized by a measuring nail in the pavement at a distance of 80 m from building.

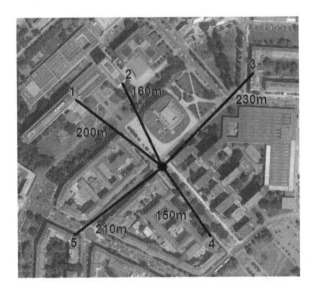

Figure 2.   Test field for horizontal directions.

Figure 3.   Points of test field for horizontal directions.

Figure 4.   Test field for vertical angles.

Figure 5.   Points of test field for vertical angles.

4   EXPERIMENTAL MEASUREMENT

Experimental measurements of all 12 total stations were the same. It was to a maximum compliance with the procedure described in ISO 17123-3. One tripod was selected for the test measurements and was always placed in the same direction on the tripod to avoid centration errors. This procedure was chosen for later testing of the horizontal angles.

Horizontal directions were measured in 4 series. One series consisted of measuring the 3 sets of horizontal directions to 5 targets with an angular misclosure. One set was measured first in face position I of the telescope and then in face position II of the telescope. According to the international standard, each series should be measured on another day and under other atmospheric conditions. Due to the number of tested instruments and time savings 2 series were measured in one day. The two series were separated by a minimum of 2 hours.

Zenith angles were measured in 4 series. One series was made by measuring 4 points in 3 sets. One set was measured first in face position I of the telescope and then in face position II of the telescope. According to the international standard, each series should be measured on another day and under other atmospheric conditions. Due to the number of tested instruments and time savings 2 series were measured in one day. The two series were separated by a minimum of 2 hours.

Measurement was conducted between November 11, 2016 and March 30, 2017. Complete testing of 12 instruments was performed in 11 days. The days for the measurement were picked when it did not rain and did not blow too much. The instrument has been conditioned for at least 30 minutes before the measurement. One series of horizontal directions lasted

30 minutes. One series of zenith angles lasted 15 minutes. If it was too windy it was not possible to measure the horizontal directions because the wind shocks influenced the compensator and adjustment of the instrument. This phenomenon was caused by placing standpoint in an urban environment in the middle of the intersection.

5 RESULTS

The measured values were processed exactly according to the procedures in International Standard ISO 17123-3. Each series was processed separately and the resulting sample deviation was calculated from 4 results of series. The standard deviations are determined by residuals as:

$$s = \sqrt{\frac{\sum\limits_{i=1}^{m}\sum r_i^2}{m \cdot (n-1) \cdot (k-1)}} = \sqrt{\frac{\sum\limits_{i=1}^{m}\sum r_i^2}{32}} \tag{1}$$

where s = final standard deviation of horizontal direction; r_i = residuals (second differences); m = number of series (4); k = number of points (5); n = number of sets (3).

For the zenith angle the resulting formula is the same with the difference that $k = 4$ and is not considered $k\text{-}1$ but only k.

The resulting standard deviations must be compared with the limit value. The null hypothesis is formulated as: Is the calculated experimental standard deviation s smaller than the value σ stated by the manufacturer or smaller than another predetermined value σ? For statistical test is used chi-squared distribution:

$$s \leq \sigma \cdot \sqrt{\frac{\chi_{1-\alpha}^2(v)}{v}} = \sigma \cdot \sqrt{\frac{\chi_{1-\alpha}^2(32)}{32}} = \sigma \cdot 1.20 \tag{2}$$

where v = degrees of freedom (32).

When considering the manufacturer's standard deviation $\sigma_{AM} = 1.5$ mgon is the limit value 1.8 mgon. From the results shown in Table 2 it is clear that all instruments meet the manufacturer's precision. Standard deviations of zenith angles would also meet instrument with a standard deviation $2''$ (0.6 mgon).

For comparison of two devices is used F-distribution and null hypothesis has the form $\sigma = \sigma'$ which consider that experimental standard deviations s and s' are obtained from two samples of measurements by different instruments.

Table 2. Final standard deviations of horizontal directions and zenith angles.

Total station	s-horizontal mgon	s-zenith mgon	Limit value mgon	s ≤ limit value
No. 1	1.17	0.62	1.8	OK/OK
No. 2	1.21	0.56	1.8	OK/OK
No. 3	1.07	0.48	1.8	OK/OK
No. 4	1.58	0.50	1.8	OK/OK
No. 5	1.16	0.47	1.8	OK/OK
No. 6	0.81	0.54	1.8	OK/OK
No. 7	0.96	0.53	1.8	OK/OK
No. 8	0.94	0.52	1.8	OK/OK
No. 9	1.32	0.53	1.8	OK/OK
No. 10	1.19	0.55	1.8	OK/OK
No. 11	1.72	0.62	1.8	OK/OK
No. 12	0.89	0.46	1.8	OK/OK

$$\frac{1}{F_{1-\alpha/2}(v,v)} \le \frac{s^2}{s'^2} \le F_{1-\alpha/2}(v,v) \quad \rightarrow \quad 0.49 \le \frac{s^2}{s'^2} \le 2.02 \tag{3}$$

where v = degrees of freedom (32); $1-\alpha$ = confidence level (0.95).

From the test of horizontal directions it was found that not all the standard deviations correspond and the instruments are therefore differently precise. The instrument No. 11 is the most different. The instrument No. 6 is the best. The zenith angles test showed that all instruments correspond.

The International Standard provides a statistical test of zenith angles for determining the magnitude of an index error (Student's t-distribution). According to this test, 11 instruments (except no. 3) suffer from a negligible index error in the range of 0.2 mgon to 4.0 mgon.

6 CONCLUSION

The paper briefly presented the procedure for testing the angular precision of the total stations. It has been verified that international standard ISO 17123-3 is suitable for common measuring procedures and provides reliable results. It is not difficult to build a test field according to this international standard and anyone can test instruments. As part of the experimental measurements a set of 12 total stations Trimble M3 was verified. For all instruments it has been confirmed that the precision of the measurement meets the precision specified by the manufacturer. Experimental measurements have shown that some instruments have better precision than manufacturer declare. It has also been shown that the instruments are very reliable even though they are commonly used in education students at the Faculty of Civil Engineering.

This work was supported by the Grant Agency of the Czech Technical University in Prague, grant No. SGS17/067/OHK1/1T/11 "Optimization of acquisition and processing of 3D data for purpose of engineering surveying, geodesy in underground spaces and laser scanning".

REFERENCES

Bručas, D. & Giniotis, V. & Petroškevičius, P. 2006. Basic construction of the flat angle calibration test bench for geodetic instruments. *Geodezija ir Kartografija* 32(3): 66–70.
Bručas, D. & Šiaudinytė, L. & Rybokas, M. & Grattan, K. 2014. Theoretical aspects of the calibration of geodetic angle measurement instrumentation. *Mechanika* 20(1): 113–117.
ČSN ISO 17123-3. 2005. *Optics and optical instruments—Field procedures for testing geodetic and surveying instruments—Part 3: Theodolites*. Český normalizační institut, Czech Republic.
Dvořáček, F. 2016. Long Range Distance Measurement with Leica AT401 Laser Tracker. In *16th International Multidisciplinary Scientific Geoconference SGEM 2016 – Informatics, Geoinformatics and Remote Sensing. Bulgaria*. vol. 2, pp 397–404, ISBN 978-619-7105-69-8.
Křemen, T. & Koska, B. 2015. Testing of Measurement Accuracy of the Total Stations. In *15th International Multidisciplinary Scientific GeoConference SGEM 2015 – Informatics, Geoinformatics and Remote Sensing. Bulgaria*. vol. 2, pp 483–489. ISBN 978-619-7105-35-3.
Kůdela, P. 2017. Testování úhlové přesnosti totálních stanic Trimble M3 – Bachelor thesis, Faculty of Civil Engineering CTU.
Lambrou, E. & Nikolitsas, K. 2015. A new method to check the angle precision of total stations. In *FIG Working Week 2015 – From the Wisdom of the Ages to the Challenges of the Modern World, Bulgaria*.
Nestorovic, Ž. 2013. On the precision of measuring horizontal directions in engineering projects. *Geonauka* 1(3): 33–40.
Rüeger, J.M. 1990. *Electronic distance measurement: an introduction*. Springer-Verlag, Germany.
Štroner, M. & Suchá, J. 2008. Ověřování souboru totálních stanic TOPCON GPT-2006 v praxi. *Stavební obzor* 17(6): 189–191.
Štroner, M. & Urban, R. & Rys, P. & Balek, J. 2014. Prague Castle Area Local Stability Determination Assessment by The Robust Transformation Method. *Acta Geodynamica et Geomaterialia* 11(4): 325–336. ISSN 1214-9705.

Historical boundary marks—an outstanding heritage

O. Dudáček & V. Čada
Department of Geomatics, Faculty of Applied Sciences, University of West Bohemia,
Pilsen, Czech Republic

ABSTRACT: In this paper, the importance of historical boundaries and boundary marks is discussed. The first part of the paper presents the Catastrum Grenzsteine project—an international project initiated by The Austrian Society for Surveying and Geoinformation (ASG) on the occasion of 200th anniversary of the establishment of the stable cadastre. The project aims on historical boundary marks in engaged states (e.g.: Czechia, Poland, Italy, Slovenia, etc.). The goal of the project is an establishment of an international database of historical boundary marks and, which is the main goal, submission of the most outstanding boundary marks for the UNESCO World Heritage title.

In the context of the Catastrum Grenzsteine project, boundary marks are presented as a cultural heritage in the second part of the paper. An outstanding value of the boundary marks in a scope of regional development, neighbours' relations and a feeling of tangible legal certainty and protection of ownership are discussed and presented in a correlation with the UNESCO World Heritage criteria. Several examples of historical boundary marks from Czechia as well as other European countries are given in this part as well.

The third part of the paper focuses on a specific importance of historical boundary marks to real estate cadastre maintenance and renewal. Since maps of the real estate cadastre of the Czech Republic (and cadastres of many other states in the area of former Austrian monarchy as well) is based on the so called stable cadastre, an exploitation of historical boundary marks as identical points for creation of transformation keys between the coordinate system of the stable cadastre and the national coordinate system is evident. Another example is an exploitation of a recognisable boundary marks during a new demarcation of boundary in terrain.

Conclusion on what has been done and proposed future steps—including development of collectors web application for boundary marks data crowdsourcing, workflow for utilisation of an open-source structure-from-motion software for 3D models of boundary marks, etc. – are given in the last part of the paper.

1   INTRODUCTION OF THE PROJECT CATASTRUM GRENZSTEINE

The project Catastrum Grenzsteine (word-to-word translation: Cadastre boundary-stones) is an international project initiated by the Austrian Society for Surveying and Geoinformation (ASG). National working groups in participating states declare their willingness to participate on the project by subscription of the so called Letter of Intent. On 22th May 2017, the Letter of Intent has been subscribed by the president of the Czech Union of Surveyors and Cartographers (CUSC). Hence, the Czech Republic has joined Austria, Hungary, Italy, Poland, Romania, Slovakia, Slovenia and Switzerland. Besides CUSC, The University of West Bohemia, Faculty of Applied Sciences, Department of Geomatics participate on the project. Also several employees of the Czech Office for Surveying, Mapping and Cadastre (COSMC) are engaged in the national working group and the president of the Office supports the idea of the project. However, the Office itself will not be engaged officially.

The main goal of the project is a submission of selected historical boundary marks and, eventually, related infrastructure (such as are trigonometric points, length bases etc.) to the UNESCO World Heritage List. There has been an idea, that the submission of such monuments of a land survey would be a perfect celebration of the occasion of the 200th anniversary of the establishment of the stable cadastre in 2017 (Stable cadastre was established in 1817 by the supreme patent on land tax and land surveying by Francis I of Austria).

Another goal of the project is to arouse a public interest in boundaries and boundary marks. For that purpose, an international database of historical (or for some other reason important) boundary marks is created. Some of the listed boundary marks are accessible from temporary webpage of the project: http://www.grenzsteine.at/.

2 HISTORICAL BOUNDARY MARKS AS A CULTURAL HERITAGE

Historical boundaries, that are mostly still valid, are unique evidences of continuity in a management of a land (Dudáček & Čada, 2017). The history of forming of boundaries has started deep in the past and requirements for demarcation have changed over time. In the past, the possession of a land was centered in hands of rulers, tribal leaders or ruler dynasties. The need of demarcation of boundaries occurred in order to allow enfeoffment of a land. Those boundaries were demarcated by natural features in terrain—mountain ranges, rivers, forests etc. An acceptable uncertainty of such boundaries was dozens of kilometers, hence, specialised technology of boundary demarcation was unnecessary.

After apanages had become hereditary property of nobility dynasties or had been donated to monastic orders, more precise demarcation of boundaries became necessary in order to avoid discords among neighbours and to support cogency of ownership.

Around the year 1270, the institution of Land Court was established. The primary reason of the establishment was an arbitration of discords among nobles. After the Ottokar II of Bohemia had been defeated (1278), Land Courts (in Prague, Brno and Olomouc) were controlled by the nobility to the 17th century. The Renewed Land Constitution (1627) returned the control over Land Courts to the Bohemian Royal Chamber (Hledíková et al. 2007). Agenda of Land Courts consisted of the arbitration of discords and criminal cases of nobility and of the surveillance over exchanges, sales, inheritances etc. of lands and other real properties. Entries related with lands or real properties were written in Domesday Books and were considered to be the strongest evidence during eventual discords.

The nobility had the right of intabulation without any conditions, while the intabulation right of towns and clergy was subject to consent of the King of Bohemia. Only several towns (Kutná Hora, České Budějovice, Pilsen and others) had the privilege that grants the right of intabulation to individual burghers of such cities. In these consequences, the importance of permanent demarcation of boundaries (of lands as well as administrative subdivisions) and of registries was increasing. Boundary marks physically present in terrain always had a higher legal relevance than the state registered in registry. Boundary marks have always been subjects of significant legal protection for that reason.

Installation and maintenance costs of such boundary marks were paid by the owner who benefited from the demarcation. Hence, a design of boundary marks was a mirror of owner's financial circumstances, too. An important attribute was also the representativeness of marks as breakpoints of boundaries.

Historical experiences have proven the importance of high quality demarcation and registration of boundaries—since the middle of the last century, it has been resigned to quality of boundary demarcation in the countries where the private ownership has been limited. A large number of boundary marks has been destroyed systematically and registration of ownership has been successively replaced by the registration of land use regardless of ownership.

More than quarter of century after 1989, the interrupted continuity of ownership registration still complicates democratisation processes in the Czech Republic. Restitution claims of eligible owners or, recently, restitution claims of churches limit the use of land, development projects, traffic infrastructure projects etc. The improvement of the real estate cadastre of the Czech Republic is, beside other reasons, complicated by the loss of historical boundary marks and by the limited accuracy of correspondent technical documentation (cadastre maps, original geodetic bases etc.).

Since the time, when accurate boundary became necessary, the quality of boundary data in registries can be only guaranteed, if land survey is based on the state of the art in sciences (math, optics, physics, mechanics and others). It is not necessary to highlight, that a quality of land survey rapidly decrease when carried out by unskilled staff.

Only boundaries that fulfill quality characteristics can provide certainty for landowners as well as authorities, which is necessary for peaceful neighbourliness, regional development, spatial planning or peaceful use of land.

The UNESCO committee has specified ten World Cultural Heritage criteria—at least one of them must be met by each cultural site being nominated for the World Cultural Heritage title (Vlčková et al. 2011). Taking the above mentioned characteristic into account, the following UNESCO criteria have been proposed as the most adequate for historical boundary marks (Ernst et al. 2017, Waldhauesl et al. 2015):

- Criterion II: "To exhibit an important interchange of human values, over a span of time or within a cultural area of the world, on developments in architecture or technology, monumental arts, town-planning or landscape design."
- Criterion IV: "To be an outstanding example of a type of building, architectural or technological ensemble or landscape which illustrates (a) significant stage(s) in human history."
- Criterion VI: "To be directly or tangibly associated with events or living traditions, with ideas, or with beliefs, with artistic and literary works of outstanding universal significance. (The Committee considers that this criterion should preferably be used in conjunction with other criteria)."

Additionally, it is believed, that boundary and boundary marks also partly conform following criteria:

- Criterion I: "To represent a masterpiece of human creative genius."
- Criterion III: "To bear a unique or at least exceptional testimony to a cultural tradition or to a civilization which is living or which has disappeared."
- Criterion V: "To be an outstanding example of a land-use which is representative of a culture or human interaction with the environment."
(Criteria have been shortened.)

3 HISTORICAL BOUNDARY MARKS DURING THE REAL ESTATE CADASTRE RENEWAL AND MAINTANANCE

As has been noted in the previous chapters, historical boundary marks are not only evidence of the past. In many cases, they are still valid breakpoints of current boundaries (Dudáček & Čada, 2017).

In the Czech Republic, the last contiguous large scale map serie based on survey in terrain has been stable cadastre. Maps of stable cadastre have been exploited as source maps for all following registries including the cadastre of lands (1931–1956). Maps of the cadastre of lands are the most often used source maps for creation of vectorised cadastral maps. However, the transformation from the coordinate systems of stable cadastre to the national coordinate system S-JTSK is needed. Once a scanned map of the cadastre of lands is transformed into coordinate System of Uniform Trigonometric Cadastral Network (S-JTSK) using global transformation key, shifts might still occur. In that case, the historical boundary marks can be used as identical points of non-residual transformation (Čada 2003).

Another example of an exploitation of historical boundary marks might be a situation, when the accuracy of a given boundary coordinates stored in real estate cadastre does not fulfill owners' requirements or when an error occurs. In that case, a reconnaissance and survey of historical boundary marks is necessary. The registered shape and position of the boundary is than corrected according to the situation in terrain.

Marking of boundaries in terrain has changed through the time. Examples of various types of marking boundaries might be for example walls or ditches along the whole length of a boundary as well as cuts in trees or stone heaps. However, the project focuses on specialised artificial boundary marks—such boundary marks are mostly sculpted from stone and are precisely fitted in place in order to ensure unequivocal definition of boundaries in terrain and thus can guarantee the stability as well as the cogency and the visibility of boundaries. Moreover, a huge number of boundary marks is decorated with, for example, coats of arms or initials of landowners and present artworks on themselves. In this section, we give two examples of outstanding boundary marks.

The boundary mark in the Figure 1 is located in the Czech Republic, more specifically near the town Vimperk in the South Bohemia. In the past, the boundary mark was a breakpoint of the boundary between Čkyně shire and Zálezly shire. Nowadays, it is still valid boundary mark of

Figure 1.   Královácký kámen boundary mark (16th century). The pictograms on the face of the mark refer to consequences of illegal shift of the mark.

Figure 2.   Photo of Ortenburger Wappenstein boundary mark (1524) by Stefan Sick (ASG; more information at wp.catastrum.eu).

the boundary between two cadastre units—Zálezly u Čkyně and Budilov. The boundary mark is called Královácký kámen. While the Czech word "kámen" means stone, the word "Královácký" refers to the word Královák, which is the denotation of an inhabitant of Královský Hvozd. Královský Hvozd was a unique subdivision of Bohemian Kingdom from approximately the 10th century to the middle of the 19th century. Inhabitants of Královský Hvozd disposed many privileges and extraordinary self-administration rights (Holý 2007).

As the municipality of Budilov village states on its website (As of June 20, 2017, http://www.obecbosice.cz/index.php?nid=3781&lid=cs&oid=5018232), the boundary mark was fitted in place probably in the 16th century. There is a ditch along the boundary near to the boundary mark. The pictograms of torture or execution instruments that have been sculpted on the boundary mark refer to consequences of a hypothetical illegal shift of the boundary mark.

In the Figure 2, there is the example of the boundary mark Ortenburger Wappenstein in Austria. This boundary mark was fitted in place in 1524 between Ortenburg and Paternion dominions in Carinthia. Nowadays, it is the boundary mark between Spittal an der Drau district and Villach District. There are coats of arms sculpted on the boundary mark, which is also denoted in the name of the mark—German word "Wappen" means a coat of arms (Ernst et al. 2017).

5 FUTURE STEPS AND RESEARCH

In the Czech Republic, the project is on its beginning. In cooperation with Austrian partners, a collector application is being developed. Once the application is finished, a Czech version will be derived and integrated into the national website of the project in order to launch a crowdsourcing of historical boundary marks data.

Next future step is a selection of outstanding boundary marks from the database created via crowdsourcing and from experts' nominations—at the moment, there are about 70 boundary marks in Czech Rep. that will be the subject of further selection. Working group then collects as much as possible additional information and prepares the nomination for national Tentative List (An inventory of properties intended to consider for a nomination). Once the property is listed in the tentative list for at least one year, a competent state authority can propose the property for the World Cultural Heritage title (Vlčková et al. 2011).

Important issue is a presentation of the project for both experts and wide public. Since a 3D visualisation is a very attractive way, a research in this topic is carried out. Because the project faces lack of resources from its very beginning, various open-source structure from motion (SfM) software is being tested. In the Figure 3, there is a screenshot of 3D model (without texture) of the boundary mark Královácký kámen mentioned in chapter 4. The model has been created using only twelve images of the boundary mark. The MeshRecon and VisualSfM software have been exploited. A quality of the model is limited by the lack of images and can be further improved.

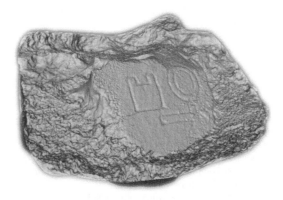

Figure 3. 3D model of the Královácký kámen boundary mark (compare with Figure 2).

The most challenging issue during the creation of 3D models is a pairwise matching among images. Paradoxically, well maintained boundary marks are often much more challenging than forgotten ones, because they are often cleaned and even painted (typically white), which makes the pairwise matching of images almost impossible. Opportunities of pre-editation of images in RAW format before conversion to JPEG format should be also tested.

Two concepts of testing have been proposed. The first one is based on comparison of volumes of tested model and that created by laser scanning. The second approach is based on subjective visual comparison of tested model and real object. Even though the second approach is unacceptable for land survey, it is believed, that it can fulfill the requirements of a web presentation for wide public.

6 CONCLUSION

This paper presents the project Catastrum Grenzsteine as a special opportunity for preservation of selected boundary marks as monuments of a land ownership. The project presents boundary marks as a testimony of an effort of our ancestors and as an outstanding technical work, which is important for peaceful neighbourliness among landowners, states etc. Being listed on the UNESCO World Cultural Heritage list would testify the importance of guaranteed boundaries for human society and therefore can promote a public perception of land surveys.

The set of appropriate UNESCO World Heritage criteria is listed in this paper and reasons, why boundary marks fulfill these criterions are given as well.

To support the theoretical motivation of the project, some examples of outstanding boundary marks including images are also given in this paper, while the last part of the paper introduces steps that are going to happen in more or less near future. The currently faced issues with a collecting of 3D data are presented, too.

ACKNOWLEDGEMENT

This contribution was supported by the Project LO1506 of the Czech Ministry of Education, Youth and Sports and the Project SGS-2016-004 Application of Mathematics and Informatics in Geomatics III.

REFERENCES

Čada, V. 2003. *Robustní metody tvorby a vedení digitálních katastrálních map v lokalitách sáhových map. (Robust methods of creation and maintenance of digital cadaster maps in localities of fathom maps).* Habilitation thesis. Prague: Czech technical university in Prague.

Dudáček, O. & Čada, V., 2017. In Ivan, I., Horák, J., Inspektor, T. (eds.), Project Catastrum Grenzsteine – State of the Art in Czechia. *Lecture Notes in Geoinformation and Cartography, Proceeding of GIS Ostrava 2017 – Dynamics in GIscience.*

Ernst, J. et al. 2017. *Network of Boundaries and Its Monuments. Draft of Tentative list submission format for transnational and transboundary future nomination.* Wien: Austrian Society for Surveying and Geoinformation.

Hledíková, Z. et al. 2007. *Dějiny správy v Českých zemích. Od počátku státu po současnost (History of the administration in the Bohemian lands. From the beginning of the state to the present).* Prague: Nakladatelství lidové noviny.

Holý, V. et al. 2007. *Královský hvozd na Šumavě před třicetiletou válkou (Královský Hvozd in Šumava before Thirty Years' War).* Domažlice: Nakladatelství Českého lesa.

Vlčková, J. et al. 2011. *Metodické zásady přípravy nominací na zápis do Seznamu světového dědictví UNESCO a zásady uchování hodnot těchto statků (Methodical principles of the preparation of UNESCO World Heritage List nominees and the principles for preserving the values of these properties).* Prague: National Heritage Institute.

Waldhauesl, P. et al. 2015. Boundary and Boundary Marks Substantive Cultural Heritage of Extensive Importance. *ISPRS Ann. Photogramm. Remote Sens. Spatial Inf. Sci.*, II-5-W3, 329–334.

Advances and Trends in Geodesy, Cartography and Geoinformatics – Molčíková et al. (Eds)
© 2018 Taylor & Francis Group, London, ISBN 978-1-138-58489-1

Multi-sensor monitoring of suspended steel bridge structure

J. Erdélyi, A. Kopáčik & P. Kyrinovič
Department of Surveying, Faculty of Civil Engineering, Slovak University of Technology in Bratislava, Bratislava, Slovakia

ABSTRACT: The weather conditions and the loading during operation cause changes in the spatial position and in the shape of bridge structures that affect static and dynamic function and reliability of these structures. Due to these facts, geodetic monitoring is integral part of bridges' diagnosis and gives important information about the current state (condition) of the structure.

To obtain complex information about the behavior of bridge structures is recommended to perform not only static deformation monitoring using conventional surveying techniques, but also dynamic deformation measurement of the monitored structure using non-conventional surveying methods.

The paper presents multi-sensor monitoring of the Liberty Bridge (Bratislava, Slovakia). The bridge is part of a cycling route between the Bratislava city district Devínska Nová Ves (Slovakia) and Schlosshof (Austria). The structure is built over an inundation area on both sides in a protected floodplain forest. The total length of the bridge structure is 525.0 m. The main structure consists of a steel structure with a triangular truss beam suspended over the river Morava and inundation bridges on both sides of the river. The measurements were focused on determination of displacements of the suspended bridge, which consists of 3 sections with spans of 30.0 m + 120.0 m + 30.0 m = 180.0 m over the river Morava.

The paper describes the experimental deformation monitoring of the bridge performed using TLS, ground-based radar interferometry and accelerometers. The procedure of the measurement, the data processing and the results of the deformation monitoring are described.

1 INTRODUCTION

Nowadays we are seeing a significant increase of the traffic intensity, which also causes increase of the operating load of the traffic infrastructure and the objects which create it. The weather conditions and the enormous traffic load causes changes in the behavior of these structures, which affect their static and dynamic functions. Due to these facts diagnosis of bridge structures including static and dynamic monitoring becomes more and more important.

The development of measuring instruments allows contactless static deformation monitoring of bridge structures. The new trends in engineering surveying show that for this purpose the technology of terrestrial laser scanning (TLS) is applicable (Zogg & Ingensand 2008), (Schneider 2006), (Schäfer et al. 2004), (Wujanz 2016). TLS allows non-contact documentation of the behavior of the monitored structure. The scan rate of current scanners (up to 1 million points per second) allows a significant reduction of the time necessary for the measurements and increases the quantity of the information obtained about the object measured. The accuracy of determination of the 3D coordinates of single measured points by currently commercially available laser scanners is several millimeters. The precision can be increased using suitable data processing, when valid assumptions about the scanned surface are available (Vosselman & Maas 2010). It can be solved by approximation of the chosen parts of the monitored object by fitting geometric primitives to point cloud. In this case, the position of the monitored point is calculated from tens or hundreds of scanned points. Using regression algorithms in combination with effective calculation software, the data processing can be significantly automated.

The knowledge of the dynamic characteristics of a bridge structure's behavior is also increasingly important. They are mainly caused by wind and moving of objects on the structure (pedestrians, cyclists, vehicles). These affect the resonant behavior of the structure, which results in the dynamic deformation of the structure. It is described by the modal characteristics of the structure's deformation (vibration modes). For the safe operation of the structure it is necessary to design these deformations by computational modelling and to monitor them during loading tests. For Structural Health Monitoring (SHM) of bridges are mainly used accelerometers, ground-based radar interferometry, GNSS, tilt sensors etc.

The technology of ground-based radar interferometry is increasingly used for the dynamic deformation monitoring of civil engineering structures (Bernardini et al. 2007), (Pieraccini et al. 2007). The radar measurements use the Stepped Frequency Continuous Wave (SF-CW) technique. This approach enables the detection of target displacements in the radar's line of sight. The advantage of radar interferometry is the determination of the displacements of the entire object measured at once. In addition to the vibration frequencies, the displacements from the direction of the line of sight can be transformed to a defined direction, e.g. vertical (vertical displacements and their frequencies).

The dynamic characteristics of the monitored objects are calculated using spectral analysis methods from the data (time series) acquired by the above-mentioned technologies. In practical applications, a finite number of the data is analyzed by the numerical method of the Fourier transformation, known as the discrete Fourier transformation (DFT). Calculation of the DFT can be realized by several algorithms. In the case of the dynamic deformation of bridges, the fast Fourier transformation (FFT) is most often used. The FFT is defined as

$$X_x(f) = \sum_{k=0}^{M} \gamma_x(k) \cdot w_x(k) \cdot e^{i2\pi f k / f_s} \tag{1}$$

where $\gamma_x(k)$ is the autocorelation function, and $w(k)$ is the spectral window function (Cooley & Tukey 1965). An alternative is the application of the Welch method, which uses the FFT algorithm too. In this case, the spectral density of the time series is computed from overlapped segments. These segments are analysed by the FFT method. The results give a smooth periodogram and greater accuracy of the frequencies determined. However, the resolution of the magnitude spectrum is unfortunately lower (Welch 1967).

Cross-spectral analysis of two time series (signals) is used for the determination of the cross-correlation and the time delay between them. It can be described as a different dynamic response to external effects (wind, pedestrians, cyclists, etc.). The cross-spectral density of two time series can be estimated by the FFT of the cross-correlation function as

$$X_{xy}(f) = \sum_{k=0}^{M} \gamma_{xy}(k) \cdot w_x(k) \cdot e^{i2\pi f k / f_s} \tag{2}$$

where $\gamma_{xy}(k)$ is the cross-corelation function, and $w(k)$ is the spectral window function (Bracewell 1965). The correlation of two time-synchronized signals at a specific period can be defined by their coherence. The significant frequencies of the signals are determined by Fisher's periodicity test. The amplitudes and the phase shifts of the signals can be estimated by the least squares method.

The paper briefly describes the experimental deformation monitoring of the Liberty Bridge (Bratislava, Slovakia) performed using TLS, ground-based radar interferometry and accelerometers.

Figure 1. Liberty Bridge, suspended structure (left) and inundation area (right).

2 CHARACTERISTICS OF THE LIBERTY BRIDGE

The Liberty Bridge is part of a cycling route between Bratislava city district Devínska Nová Ves (Slovakia) and Schlosshof (Austria). It crosses the river Morava at the river kilometer 4.31, where a transverse cycling route, a stagnant pool of the river Stará mláka, the ruins of the old bridge, and a border fortification bunker are located. The bridge is built over an inundation area on both sides in a protected floodplain forest (Fig. 1). The total length of the bridge structure is 525.0 m (Agócs & Vanko 2011).

The substructure consists of reinforced-concrete pillars in which the supports of the main structure are anchored. The main structure consists of a steel structure with a triangular truss beam suspended over the river Morava and inundation bridges on both sides of the river.

The measurements were focused on determination the static and dynamic displacements of the suspended bridge, which consists of 3 sections with spans of 30.0 m + 120.0 m + 30.0 m = 180.0 m over the river Morava. The reinforcing girder is a tubular triangle shaped girder with an orthotropic deck. The middle section has the shape of a circular arc with a radius of 376.35 m. The deck consists of a metal plate, steel girders positioned in transverse direction, and longitudinal reinforcements. The cross slope of the deck is 2% from the longitudinal axis of the bridge to the edges; the clearance width is equal to the width of the 4.0 m traffic lane. The structure of the main section is suspended on four pylons, which are designed as dual—hinged rectangular frames. The diameter of the pylons is 0.914 m; their height is 17.7 m (Fig. 1).

3 DEFORMATION MONITORING

The monitoring using TLS was performed in 3 measurement epochs using Leica ScanStation 2. The bottom side of the middle section of the suspended structure was scanned from a single position of the scanner. The reference network consists of four control points. Since the bridge is built in a natural reservation, there are no possibilities to make observation sites. Due to this restriction two of the control points are stabilized on the base of the pillars on the Slovak side by metallic fasteners, and two of them are on the points of the original setting-out network of the bridge and stabilized by observational pillars. All the control points were signalized by Leica HDS targets. The data obtained by the TLS were transformed to the local coordinate system of the bridge defined by the control points.

The main task of the data processing was modelling the position of the monitored points (using orthogonal planes). These are positioned on the bottom side of the transverse girders between the diagonal reinforcements of the supporting girder on the both sides of the bridge. The total number of monitored points was 46. The vertical displacements of the points were determined as the difference between heights of these points in each epoch. The height of the points was calculated using orthogonal regression. During the data processing of the initial measurement square fences 75 mm × 75 mm were defined on the bottom side of the girders. These fences define approximately the same set of points in each measurement epoch.

The figure (Fig. 2) shows the displacements of the monitored points selected. The measurements show the displacement of all the observed points except for the points on the ends of the suspended structure. These points have not changed their position, because the structure is anchored to the supporting structure at these parts. The standard deviation of the displacements was calculated using uncertainty propagation law from the vertical component of the transformation error and the standard deviation of the regression planes as:

$$\sigma_{Z_P} = \sqrt{\sigma_{T_Z}^2 + \sigma_\rho^2} \qquad (3)$$

where: σ_{T_Z} is the vertical component of the transformation error and σ_ρ is the standard deviation of the calculated regression plane (calculated from orthogonal distances of the points of point cloud from the plane). The accuracy of the transformation is given by the differences (ΔX, ΔY, ΔZ) between the identical reference points. The vertical component σ_{T_Z} was calculated as a quadratic mean of differences ΔZ between the common reference points.

The standard deviation of the displacements varies from 1.3 mm to 1.8 mm. The displacements towards the center of the bridge are increasing and have negative values. In the middle of the bridge they reach values of −13 mm. This is partly caused by the lower temperature of the structure in the control epoch of the measurement and partly by the load of a 10 cm layer of fallen snow.

For dynamic deformation monitoring of the bridge two HBM B12/200 one-axial accelerometers, which are supported by an HBM Spider 8 A/D transducer and a ground-based IBIS-S interferometric radar was used.

The accelerometers measure acceleration in vertical direction. These inductive sensors have an operating frequency of up to 200 Hz and measuring range of up to 200 m.s^{-2}. The accuracy of the sensors is defined by a relative error of up to ±2%. The sensors were positioned in the middle of the suspended structure and at the anchorage of the suspension cables (Fig. 3).

The radar measures dynamic displacements by comparing the phase shifts of reflected radar waves collected at the same time intervals. The displacement is measured in a radial direction (line of sight). The minimal range resolution of the radar used is 0.5 m. The accuracy

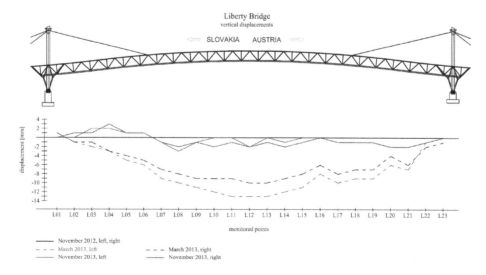

Figure 2.   Vertical displacements of the suspended structure.

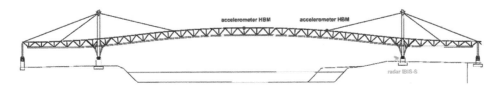

Figure 3. Position of the radar and the accelerometers.

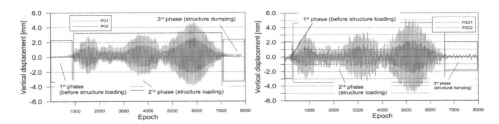

Figure 4. Vertical displacements determined by radar (left) and accelerometers (right), 1 person jumping.

of the measured displacements is at the level of 0.01 mm, but it depends on the range and the quality of the reflected signal. The measurement and the data registration is managed by the IBIS-S operational software installed in a PC (notebook).

Four dynamic loading epochs were defined as follows: 1. - without loading, 2. - 1 person, walking, 3. - 1 person running, 4. - 1 person jumping at the center of the structure. Each epoch was performed in 3 phases. In the first phase, the measurements were started before the loading of the structure. The second phase continued during the loading and the last phase was realized after the loading of the structure during the damping of the structure. Each epoch lasted approximately 2 minutes. The frequency of the data registration by the accelerometers and ground-based radar were realized on the level of 100 Hz due to the requirements to achieve higher accuracy of the relative displacements and the occurrence of significant frequencies of structural deformations, which were higher than 10 Hz.

The relative displacements using the accelerometers was calculated by double integration of the accelerations measured. The accelerometer drift and integration errors were eliminated by a Butterworth high-pass filter with a cut-off frequency at the level of 0.5 Hz. This filter attenuates the magnitude of the spectrum at the frequency of 1 Hz by 0.7%, which has no significant influence on the displacements determined. The filter was applied before and after the first integration of the velocities.

The effect of pedestrians walking has a minimum influence on the vertical displacements. The rapid movement of pedestrians affects the maximum vertical displacements twice more than during the loading by a pedestrian's walking. The harmonic jumping of pedestrians affects the maximum displacements at a level of around 2.50 mm (P01) and 4.85 mm at the center (P02) of the structure.

Before the spectral estimation of the accelerations, each measured time series of the accelerations was filtered by the Butterworth high-pass filter with a cut-off frequency of 0.1 Hz. This filter attenuates the signal amplitudes with a 1 Hz frequency on the level of 0.2%. This has a minimum influence on the estimation of the expected dominant frequencies of the structural deformations. Determining the natural frequencies of the structural deformations at each measured point (P01 and P02) was realized by auto-spectral analysis using the FFT method.

Cross-spectral analysis of the time series measured by the acceleration sensors and interferometric radar at the points P01 and P02 was realized by the FFT method too. The estimation parameters were the same as during the auto-spectral estimation at these points. In the next step, the phase shifts of both signals and their coherence were determined.

During the measurement without the structure's load has been determined frequency of deformation at the level of 1.52 Hz only by radar measurements. Signals from both measuring points are low coherent with relatively high phase delay at the level of around 65.0°. The dominant frequencies of the deformations which were determined during the walking of 1 person corresponds with the 22nd vibration mode of the structure (from the FEM model). The estimated frequencies approximately at the level of 2.10 Hz corresponds with the frequency of the pedestrian's steps during standard walking. The phase shift at the level around 23.0° is caused by a short delay in the structure's response at the points P01 and P02, which were influenced by pedestrian walking. The 3rd loading epoch shows the resonant oscillation of the structure affected by running in the range from 2.50 Hz to 3.00 Hz, which corresponds to the frequency of the impact of feet on the structure during running. The signals at the 44th vibration mode at the level around 3.80 Hz have a minimum phase delay in a range from 0° to 3°. The 4th load was realized during synchronized jumping by one person at the center of the structure. The structural oscillation at the frequency level of 1.81 Hz and the minimum phase shift are affected just by this activity. The estimated frequency is similar to the 22nd natural frequency defined by FEM model.

4 CONCLUSIONS

The weather conditions and the loading during operation cause changes in the spatial position and in the shape of bridge structures that affect static and dynamic function and reliability of these structures. Due to these facts, geodetic monitoring is integral part of bridges' diagnosis and gives important information about the current state (condition) of the structure. To obtain complex information about the behavior of bridge structures is recommended to perform not only static deformation monitoring using conventional surveying techniques, but also dynamic deformation measurement of the monitored structure using non-conventional surveying methods.

The paper deals with the multi-sensor deformation monitoring of the Liberty Bridge by non-conventional surveying methods: terrestrial laser scanning, accelerometers and ground-based radar interferometry. The basic principles of deformation monitoring using the mentioned methods are described. The paper brings basic recommendations of methodologies used, however the measurement with the above-mentioned methods have to be modified according to the specifications of the monitored object. For deformation monitoring using TLS is recommended to use suitable data processing, e.g. regression models to increase the precision of the results (in final the accuracy also). For dynamic deformation monitoring is important to collect data with higher ratio than the frequencies wanted to be determined.

ACKNOWLEDGEMENT

This article was created with the support of the Ministry of Education, Science, Research and Sport of the Slovak Republic within the Research and Development Operational Programme for the project "University Science Park of STU Bratislava", ITMS 26240220084, co-funded by the European Regional Development Fund.

REFERENCES

Agócs, Z. & Vanko, M. 2011. "Devínska Nová Ves—Schloshof Cycling bridge. Cyklomost Devínska Nová Ves—Schlosshof", Technical Documentation. Bratislava: INGSTEEL, spol.s.r.o. 2011. p. 29.
Bernardini, G., De Pasquale, G., Bicci, A., Marra, M., Coppi, F. & Ricci, P. 2007. Microwave interferometer for ambient vibration measurements on civil engineering structures: 1. Principles of the radar technique and laboratory tests. EVACES '07, 2007.

Bracewell, R. 1956. Pentagram Notation for Cross Correlation. The Fourier Transform and Its Applications. New York: McGraw-Hill, pp. 46–243, 1965.

Cooley, J.W. & Tukey, J.W. 1965. An algorithm for the machine calculation of complex Fourier series. Mathematic Computation. 19 (90). pp. 297–301.

Pieraccini, M., Parrini, F., Fratini, M., Atzeni, C., Spinelli, P. & Micheloni. M. 2007. Static and dynamic testing of bridges through microwave interferometry. NDT&E Int, 40 (2007), pp. 208–214.

Schäfer, T., Weber, T., Kyrinovič, P. & Zámečníková, M. 2004. Deformation measurement using Terrestrial Laser Scanning at the Hydropower Station of Gabčíkovo. In: INGEO 2004 and FIG Regional Central and Eastern European Conference on Engineering Surveying [CD-ROM]. Bratislava: KGDE SvF STU, 2004, 10 p. ISBN 87-90907-34-05.

Schneider, D. 2006. Terrestrial Laser Scanning for Area Based Deformation Analysis of Towers and Water Dams. In: Proceedings of the 3rd IAG Symposium on Geodesy for Geotechnical and Structural okoEngineering and 12th FIG Symposium on Deformation Measurements [CD-ROM]. Baden: TU Wien, 2006, 10 p.

Vosselman, G. & Maas, H.G. 2010. Airborne and Terrestrial Laser Scanning. Dunbeath: Whittles Publishing, 2010. 318 s. ISBN 978-1904445-87-6.

Welch, P.D. 1967. The Use of Fast Fourier Transform for the Estimation of Power Spectra: A Method Based on Time Averaging Over Short, Modified Periodograms. IEEE Transactions on Audio Electroacoustics, pp. 70–73.

Wujanz, D. 2016. Terrestrial Laser Scanning for Geodetic Deformation Monitoring. Dissertation, TU Berlin.

Zogg, H.M. & Ingensand, H. 2008. Terrestrial Laser Scanning for Deformation Monitoring—Load Tests on the Felsenau Viaduct (CH). In: The International Archives of the Photogrammetry, remote Sensing and Spatial Information Sciences, Bejing, p. 555–562. http://www.isprs.org/proceedings/XXXVII/congress/5_pdf/97.pdf Accessed 10 July 2015.

Advances and Trends in Geodesy, Cartography and Geoinformatics – Molčíková et al. (Eds)
© 2018 Taylor & Francis Group, London, ISBN 978-1-138-58489-1

Pedestrian positioning using smartphones in building with atypical geometry

Ľ. Erdélyiová, P. Kajánek & A. Kopáčik
Department of Surveying, Slovak University of Technology in Bratislava, Bratislava, Slovakia

ABSTRACT: The article deals with the application of inertial sensors (accelerometers and gyroscopes built in smartphones) for pedestrian positioning in the indoor environment. The problem of using inertial sensors results from their functional principle, which is based on the integration of inertial measurements. A secondary product of integration is the rapid accumulation of relatively small errors of inertial sensors in the actual position and orientation of pedestrian. To eliminate the systematic errors of the accelerometers, model of data processing uses adaptive estimate of the step length. This algorithm uses the actual amplitude of acceleration and the walking frequency to estimate the step length.

The often-applied approach to eliminate the systematic errors of gyroscopes is the approximation of direction of pedestrian motion in four main directions, but this solution assumes the rectangular geometry of buildings. In order to eliminate this limitation, the article proposes new model which is extended by algorithm for identification rooms and a combined calculation of orientation to eliminate the systematic errors of gyroscopes. The article describes the algorithm for identifying the room where the pedestrian is located. On the base of the identified room, the model identifies the room geometry (from floor map of the room) and the corresponding solution (with or without approximation of orientation). In the case of room with atypical geometry, the orientation calculated without approximation and the map matching algorithm is used for verification and correction of the calculated position of pedestrian based on map information of the selected room.

1 INTRODUCTION

The development of Micro Electro Mechanical Systems (MEMS) technology enabled the production of small low-cost sensors with low power requirements, which significantly expanded their applications (navigation, healthcare, gaming and others). Smartphone built-in inertial sensors are easy accessible, which made them possible to use for pedestrian navigation in the indoor environment.

The principle of positioning using inertial measurement system IMS is based on the integration of inertial measurements (acceleration and angular rate) to translation movement of IMS and change of orientation of IMS (pedestrian). Integration of inertial measurement causes rapid accumulation of systematic errors in final position and orientation of IMS, which leads to an increase in position error with time of measurement (Groves, 2008).

The error model of the sensor contains, in addition to the constant components of systematic errors, which can be eliminated by calibration of the sensor (El-Diasty, 2008), also the dynamic part of systematic errors DPSE, which changes during the measurement. We cannot remove the dynamic component of systematic errors by calibration, so we need to design a processing procedure that will ensure that it is effectively eliminated in the processing process.

Nowadays, indoor navigation is actual for many scientist, who develop their own navigation system. Navigation system based on using IMS is suitable for indoor navigation, because measure actual position and orientation of user, which is independent from external signals (Kopáčik et al 2015). Another possibility is a system based on WLAN (Puertolas-Montañez et al. 2013), whose advantage is opportunity to use existing WLAN network. Navigation system based on Bluetooth (Hallberg, 2003) was originally designed for the short-range connection for personal devices but its utilization can be applied also in the methods of indoor navigation based on the triangulation method using received signal strength. Another solution is to use UWB (ultra wide band) (Renaudin, 2007) when the radio signals penetrate into buildings also through a very full environment. Often used method is also positioning by ultrasound (Sugimoto, 2009), RFID (radio frequency identification) (Nakamori et al., 2012).

2 ELIMINATION OF SYSTEMATIC ERRORS OF ACCELEROMETERS

In order to eliminate the systematic errors of accelerometers, zero velocity update ZUPT is applied. The principle of ZUPT is to apply a condition of zero-speed at a time when the IMS is stable. Based on this condition, the error's component of the calculated velocity is modeled as a linear function defined (calculated by integration of an acceleration) at the moment of starting and stopping of the IMS (or at the moment when the foot touches the ground). Due to the short time intervals between applying the zero-speed condition, this method effectively eliminates the error component of the inertial sensor (accelerometer). The disadvantage of this method is the condition for the location of the IMS that should be placed on the foot of the pedestrian (Jiménez, 2011). Since IMS (smartphone) placed in pedestrian's hand fulfills the condition of zero-speed only when pedestrian starts and stops (not at the moment when foot touches the ground), the time interval between pedestrian stops gets longer. Long time interval between pedestrian stops have negative effect on elimination of DPSE of accelerometer.

In order to exclude the conditions for the location of the IMS, the step detection method was used. The principle of this method is in the periodic movement of IMS, which is generated by the periodic motion of two legs, which can be described as inverted pendulum. Step is defined as one period of periodic motion. Individual steps are identified using zero-crossing method to measure the norm of acceleration (resulting acceleration vector). Each step consists of three zero-crossings (Fig. 1).

The step length varies during the walk according to the environment (stairs, obstacles, doors, floor surface) and walking speed. The change of step length is then reflected in measured acceleration. The actual step length is calculated using adaptive step length algorithm, where the actual step length is a function of the walking frequency f and the average amplitude of the norm of acceleration v.

$$l = \alpha \cdot f + \beta \cdot v + \gamma \tag{1}$$

$$f = \frac{1}{t_k - t_{k-1}} \tag{2}$$

$$v = \frac{1}{n-1} \sum_{k=1}^{n} \left(\|a_k\| - \overline{\|a\|} \right)^2 \tag{3}$$

where: t = time of step detection; $\|a\|$ = norm of acceleration; $\overline{\|a\|}$ = average value of norm of acceleration. A more detailed description of this method and results of experiments are presented in (Kopačik et al., 2015).

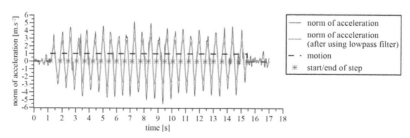

Figure 1.   Step detection based on zero-crossing method.

3 ELIMINATION OF SYSTEMATIC ERRORS OF GYROSCOPES

In order to eliminate the systematic errors of gyroscopes, it is not possible to use the periodic character of pedestrian walk. A change of pedestrian's orientation is calculated by integrating the angular velocity measured by gyroscope, whose sensitive axis is perpendicular to the plane of motion. Due to the design of the smartphone and its orientation in the pedestrian's hand (horizontal position—the pedestrian looks alternately on the display of smartphone and on the floor), we calculate the orientation of the pedestrian movement (or azimuth) from gyroscope measurements in the direction of the Z axis.

The error component generated by the integration of angular velocity has a linear trend (Fig. 2 – blue line), that corresponds to the DPSE of the gyroscopes. The first step to eliminating DPSE of the gyroscopes is to identify the drift in calculated orientation at the beginning of the measurement. We assume that at the beginning of the measurement the IMS is stable. The drift is modeled as a linear function that is defined on the basis of regression analysis. The result of this process is represented by green line in Figure 2.

During the measurement DPSE changes, therefore the use of one function characterizing the drift at the beginning of the measurement leads to an insufficient elimination of the error component represented in the calculated orientation.

Another approach to eliminating systematic errors of gyroscopes is approximation of pedestrian's orientation into four main directions. This approach is based on the assumption of rectangular geometry of buildings. The principle of the algorithm for identifying the main direction of pedestrian's movement lies in a series of conditions for the current azimuth of pedestrian movement. Using this approach for calculating pedestrian's orientation good results can be achieved in situations when the pedestrian moves through narrow corridors with rectangular geometry which restrict orientation of pedestrian's movement. The disadvantage of this approach is large spacious buildings with atypical geometry (vestibules, atriums).

3.1 Combined approach to eliminating DPSE of gyroscopes

The approximation of pedestrian orientation to the main directions effectively eliminates the systematic errors of gyroscopes but this solution fails in rooms with atypical geometry. For this reason, model of data processing was designed (Fig. 3).

This model uses a combined approach for calculation of orientation based on the room identification. Based on the identified room, the model selects an approach for calculation of the orientation (with or without approximation of orientation).

Room identification (Fig. 4) is realized in time t when the pedestrian passes through the door of the room. In the first step, proposed algorithm finds the intersection of the pedestrian trajectory with the door. The algorithm identifies the room based on the identifier. In our case the identifier is number of room stored in matrix A (or matrix B). Two matrices (A, B) are used because two different situations can occur when pedestrian passes through the door (pedestrian enters the room or pedestrian leaves the room). Storing of an identifier in the matrix (row and column) will be performed based on the coordinates of the center of the door (X_T, Y_T). The identification of the correct room from a pair of matrix is based on the current orientation of pedestrian's movement (Fig. 5).

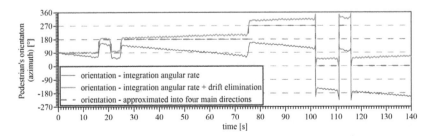

Figure 2. Elimination of the drift identified in the calculated orientation.

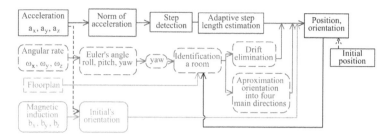

Figure 3.   Scheme of the proposed model with combined approach to calculate orientation.

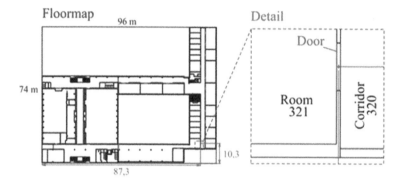

Figure 4.   The principle of algorithm for identification a room.

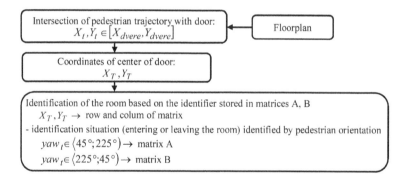

Figure 5.   Scheme of an algorithm for identification a room.

Note The room identifier 321 shown in the detail will be stored in matrix $A_{87,10}$, in matrix B at the same position $B_{87,10}$, will be room identifier 320 (number for corridor)

4   EXPERIMENTAL MEASUREMENT AND EVALUATION OF RESULTS

In order to verify the proposed model, an experimental measurement was carried out. During the experiment, the pedestrian passed along a predefined trajectory, which also passed through rooms with atypical geometry. Trajectory was defined by 12 measured points **MP**. Experiment was realized with smartphone Samsung Galaxy S4 with built-in IMS, which was placed in the pedestrian's hand. The smartphone features triaxial accelerometer, triaxial gyroscope and triaxial magnetometer. On the basis of inertial measurements, the pedestrian's trajectory was calculated using the proposed model (Fig. 3), where the following approach to calculate the orientation of pedestrian movement was used:

- Integration of angular velocity with drift elimination.
- Orientation of pedestrian movement defined by magnetic azimuth.
- Approximation of the pedestrian orientation into the four main directions.
- Combined approach to eliminating DPSE gyroscopes (combination of the first and the third approach).

The comparison of the efficiency of the proposed model with other approaches to the elimination of systematic errors of gyroscopes and their influence on the accuracy in position determination was based on positional deviations at observed points of pedestrian trajectory (Fig. 7).

From the results of the experimental measurements we can see that the highest positional deviations were achieved using a model in which orientation drift was parametrized on the beginning of the measurement, where smartphone was stable. Rapid accumulation of errors in the orientation (orientation drift 0.16°/s – Fig. 6) resulting from the insufficient elimination of DPSE leads to an increasing error in position determination. Better results were achieved using the model in which the orientation of pedestrian movement was defined by the magnetic azimuth calculated on the basis of the magnetic intensity measured by magnetometer. The advantage of magnetometers is the fact that based on magnetic induction, we can calculate an absolute orientation that is not burdened by an error from previous measurements. On the other hand, their disadvantage is lower accuracy in orientation determination and high sensitivity to changes of magnetic field generated around robust metal structures (Fig. 6) and electronic devices (Ilyas, 2016).

When pedestrians walk in buildings with rectangular geometry, the best model is the approximation of the orientation into the four main directions. But this model fails when pedestrians move in spaces with atypical geometry. The solution to this problem is model with combined approach to calculation orientation that achieves results comparable with the previous model. If we include correction of a pedestrian position in processing, based on the center the door center (the pedestrian position will be corrected to the center of the door), we will achieve a further increase in position accuracy.

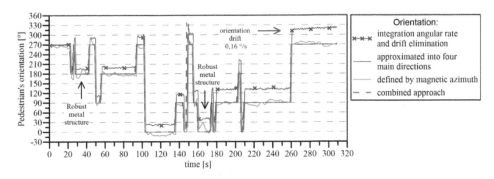

Figure 6.   Pedestrian orientation calculated by individual solution variants.

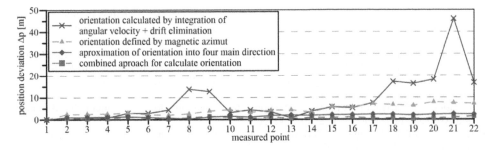

Figure 7.   Comparison of selected approaches for elimination DSCH gyroscope based on position deviation.

5 CONCLUSION

The development of MEMS technology has contributed to the production of small and low-cost IMS, which led to the expansion of their application's area. In addition to other applications, they have a significant use in pedestrian indoor navigation.

IMS allows monitoring of pedestrian movement in three-dimensional space independently of external sources and in areas where GNSS signal is not available. On the other hand, functional principle of the IMS based on the integration of inertial measurements causes a rapid accumulation of systematic errors of inertial measurement in actual position and orientation. The actual position error is generated by the systematic errors of accelerometers and gyroscopes.

To eliminate the systematic errors of the accelerometer, the proposed model uses a step detection method complemented by an adaptive step length algorithm which defines step length based on walking frequency and average amplitude of acceleration.

The paper is focused mainly on the elimination of systematic errors of gyroscopes. For this purpose, combined approach to calculate the pedestrian orientation was used. This approach combines the calculation of orientation with subsequent approximation into the four main directions in rooms with rectangular geometry and calculation of orientation without subsequent approximation in rooms with atypical geometry. In rectangular buildings, the approximation of orientation to the main directions allows to effectively eliminate the systematic errors of gyroscopes, but fails in spaces with atypical geometry. In order to identify the geometry of the building, a room identification algorithm was designed to identify the room where the pedestrian is currently located. Another benefit of the proposed model is the possibility to correct the pedestrian's position at the time of passing through the door when model fits pedestrian position to center of the door.

ACKNOWLEDGEMENT

"This publication was supported by STU Foundation for Talent Development—Scholarship of the Slavoj's family."

REFERENCES

El-Diasty, M. 2008. Calibration and stochastic modeling of inertial navigation sensor errors. 2008. Journal of Global Positioning Systems. p. 170–182.

Groves, P. D. 2008. GNSS Technology and Aplications Series, Principles of GNSS, Inertial and Multisensor Integrated Navigation Systems, London: Artech House, 2008, 552 s., ISBN-13:978-1-58053-255-6.

Hallberg, J., Nilsson, M., Synnes, K. 2003. Positioning with Bluetooth, Proceedings, 10th International Conference on Telecommunications (ICT 2003), 23 February – 1 March 2003, Vol. 2, 954–958.

Ilyas, M., et al. 2016. Drift reduction in pedestrian navigation system by exploiting motion constraints and magnetic field. In: Sensors, 2016, volume 16.9: 25 p.

Jiménez, A.R., et al. 2011. PDR with a foot-mounted IMU and ramp detection. In: Sensors, 2011, volume 11.10, p. 9393–9410.

Kopáčik, A., Ilkovičová, Ľ., Kajánek, P. 2015. Pedestrian Trajectory Determination in Indoor Environments. In: FIG Working Week 2015, technical programme and proceedings. Sofia, Bulgaria, 17. – 21. 5. 2015. – 1. edition. - ISBN 978-87-92853-35-6, 15 p.

Nakamori, E., Tsukuda, D., Fujimoto, M., Oda, Y., Wada, T., Okada, H. 2012. A New Indoor Position Estimation Method of RFID Tags for Continuous Moving Navigation Systems, 2012 International Conference on Indoor Positioning and Indoor Navigation, 13–15 November 2012.

Puertolas-Montañez, J.A., Mendoza-Rodriguez, A., Sanz-Prieto, I. 2013. Smart Indoor Positioning/Location and Navigation: A Lightweight Approach, International Journal of Interactive Multimedia and Artificial Intelligence, Vol. 2, No. 2, 43–50.

Renaudin, V., Yalak, O., Tomé, P., Merminod, B. 2007. Indoor Navigation of Emergency Agents, European Journal of Navigation, Vol. 5, No. 3, 36–45.

Sugimoto, M., et al. 2009. An Ultrasonic 3D Positioning System Using a Single Compact Receiver Unit. In: Location and Context Awareness. Springer Berlin Heidelberg, 2009. p. 240–253.

Advances and Trends in Geodesy, Cartography and Geoinformatics – Molčíková et al. (Eds)
© 2018 Taylor & Francis Group, London, ISBN 978-1-138-58489-1

Landslide movements analysis: The case of Kadzielnia in Kielce

P. Frąckiewicz, K. Krawczyk & J. Szewczyk
Geomatic and Energy Engineering, Faculty of Environmental, Kielce University of Technology, Kielce, Poland

ABSTRACT: The application of comprehensive methods of landslide movement monitoring (using scanning tacheometry, laser scanning, photogrammetry—including the use of drones) allows the collection of copious quantities of measurement data. While processing the material one should develop internally coherent and complete data using all the obtained information. The landslide monitoring in the area of Kadzielnia in Kielce has been performed for the last two years. The paper presents the methods of processing, its results with a particular emphasis put on solutions based on statistics and geostatistics. These methods allow the assessment of the risk likelihood and the ratio of random and non-random factors affecting the distribution of changes in the landslide slope shape. The obtained results should be considered as provisional. At the same time, they indicate the advisability of combining different methods of observation and analysis to obtain practical, useful results which are relevant to the protection of the environment and its qualities.

1 INTRODUCTION

Landslides are defined as a displacement of rock masses, formed in the conditions of constant contact with the ground, mainly due to the action of gravity along the slide surface. Disrupting the balance of the two forces: the one maintaining the slope (related to the strength of the formation) and the other pressing the slope cause the occurrence of landslides and other destructive mass movements.

One of the areas with intensive landslide movements is Kadzielnia—a strict reserve of inanimate nature in the center of Kielce, located in the former site of Devonian limestone mine operating from the 17th century until 1962. The monitored area of stricte reserve is 0.6 ha, the area (partially monitoring) of all object –13.5 ha, the highest elevation reaches 295 m.a.s.l., the height difference is up to 53 m.

2 LANDSLIDE OBSERVATION METHODOLOGY

Since 2014 certain selected landslides in Kadzielnia have been monitored within the cooperation between the Faculty of Environmental, Geomatic and Energy Engineering of the Kielce University of Technology and Kielce Geopark.

Monitoring of this type of landslides involves the use of a number of surveying methods: point methods, area method or hybrid methods, and also subsurface methods. In the case of Kadzielnia landslides the investigation primarily involved area methods. The following monitoring solutions were applied within the area methods:

1. Laser scanning of the landslide surface using the TOPCON scanning total station Quick Station 1A (18 observation cycles at different landslides; May–June 2014, October–November 2014, June–July and October–November 2015 and November 2016). The monitoring was performed in regular squares (with sides of about 0.50 m) or rectangles (with sides 1.0 and 1.5 m), or profiles distant from each other by about 2–4.5 m. The number of

points for which the coordinates were determined ranged from 600 to 2200 on individual landslides. The value of maximum margin of error (3σ; crucial to recognize the significance of the changes on the surface of landslides) was assumed as 0.10 m.

2. Laser scanning of the landslide surface (four monitoring cycles in September 2014 and in November 2016) using OPTECH 3D laser scanner ILRIS (in 2014) and STONEX X300 (in 2016). The number of points in the cloud ranged from 10 to 21 million.

3. Taking photogrammetric images (1 measurement in October 2016) using Phantom 3 Professional drone.

3 METHODS OF LANDSLIDE MOVEMENT ANALYSIS

The purpose of monitoring landslide surfaces is to identify changes occurring between observation cycles and consequently to determine the magnitude of displacements and losses, to identify active parts of the landslide and possible threats in this respect (for taking preventive measures).

The resulting data are commonly only an approximation of the real measured value. The point analysis of the landslide provides the information on the displacement in the spatial X,Y,Z system. The displacements are presented in the form of vectors (u,v,w). The application of this method requires stabilized measurement points. The area methods eliminate certain sources of inaccuracy owing to the acquisition of very large sets of points. The coordinates are determined with less accuracy, but the surface is usually covered with a regular dense grid, which allows the depiction of the landslide changes. The evaluation of the surface profiles is the major method of the analysis of the models developed on the basis of such profiles, and also the assessment of the gradient change or the development of the landslide plane model limited by the constant plane to determine the volumetric changes of the solid.

3.1 *Graphic comparison of landslide surfaces monitored in successive measurement cycles and the visualization of results*

Graphic presentation of results facilitates the interpretation and depicts the landslide changes in a clear, understandable manner. The obtained data should be appropriately prepared. The following programmes are applied to do this task: WinKalk 4.05, MicroMap 5.50, Quantum GIS 2.8.1-Wien, Surfer 11 (Duma et al., 2016; Frąckiewicz et al., 2016).

3.2 *Statistical analysis of monitoring results*

The statistical development of measurements aims to provide quantitative presentation of changes in the area. Three statistical methods of analysis have been applied: studying the volumetric change, the analysis of the slope gradient change and the section method in the areas of qualitative changes. Additionally, the Surfer programme permits a geostatistical evaluation by means of variograms. The case relates to a 3-fold measurement of the landslide surface (denoted as: reference, first and second measurement).

3.2.1 *Volumetric changes*

The volumetric evaluation was obtained by using the Volume functionality on the GRID. The generated models can be compared with each other, specifying which of the two represents the upper and which the lower plane. In the resultant report one obtains the basic data for both areas, positive volumes (included between the upper and the lower surfaces when the upper is located above the lower one) and negative volumes (included between the upper and lower surfaces, the upper is below the lower). In addition, it contains the information on the projected surface (Planar Areas) on the horizontal plane and the magnitude of the area measured along the slope (Surface Areas). Table 1 contains all the volume comparisons (Frąckiewicz et al., 2016).

The Volume functionality in the Surfer programme permits calculating the volume not only with regard to a specified surface, but also with respect to a defined level on the Z direction. The volume of particular solids on a number of levels was calculated (Frąckiewicz et al., 2016). The volumes are shown in Table 2.

3.2.2 *Changes in the slope inclination*
The slope inclination was determinated in the Surfer programme. The obtained report includes the basic data with reference to the object and different inclination values from the minimum ones to average, median and maximum. The angles of inclination from the horizontal were determined for all the cases with the values being converted into the decline percentage (Frąckiewicz et al., 2016). Table 3 includes all the calculated inclinations.

3.2.3 *Cross section method*
This method permits a linear or point presentation of a plane in a specified location. It can be used to determine rock mass gain or loss in the places of anticipated changes. The search for such sites occurs as a result of analyzing images with the given geographical reference or by analyzing changes in the course of contour lines on a map.

Table 1. Calculated changes in the volume between surfaces (for laser scanning measurements).

Upper surface	Lower surface	Positive volume [m³]	Negative volume [m³]	Positive surface measured along the slope [m²]	Negative surface measured along the slope [m²]
1st measurement	reference measurement	540.3	14.4	546.5	59.0
2nd measurement	reference measurement	355.5	21.5	456.5	84.2
2nd measurement	1st measurement	15.7	187.8	82.2	373.3

Table 2. Calculated changes in the surface volume at different levels (for laser scanning measurements).

Shear level [m a.s.l.]	Volume from the reference measurement above shear surface [m³]	Volume from the first measurement above shear surface [m³]	Volume from the second measurement above shear surface [m³]	Difference [5] = [3] − [2]	Difference [6] = [4] − [3]
[1]	[2]	[3]	[4]	[5]	[6]
270	109.5	103.8	111.1	−5.7	7.3
265	656.0	630.3	639.1	−25.7	8.8
260	1730.1	1693.0	1720.3	−37.1	27.3
255	3116.0	3071.6	3117.6	−44.4	46.0
250	4771.3	4727.4	4794.7	−43.9	67.3
248	5555.2	5479.4	5558.4	−75.8	79.0

Table 3. Calculated changes in the slope surface inclination (for laser scanning measurements).

Measurement	Min. inclination [°]	Min. inclination [%]	Max. inclination [°]	Max. inclination [%]	Average inclination [°]	Average inclination [%]	Median [°]	Median [%]
reference measurement	7.16	12.5	83.16	99.3	62.26	88.5	68.38	93.0
1st measurement	2.51	4.4	83.75	99.4	62.46	88.7	68.62	93.1
2nd measurement	7.97	13.9	83.19	99.3	62.30	88.5	68.27	92.9

The contour analysis (Frąckiewicz et al., 2016) allowed the selection of 21 sections with the biggest changes. The sections were compared with the calibrated raster, the sections in which the differences were due to the vegetation measurement were omitted. The digitization functionality (digitize) was used to draw the cutting planes at selected locations. The slicing functionality (slice) from a selected grid creates a profile.

3.3 *Geostatistical analysis of the monitoring results*

The study of the landslide surface involves monitoring observations that are dependent on each other due to the quasi-continuous surface. The variability describing the deformation of the area surface can be random or non-random; surface deformations can also be caused by one or several reasons (Mucha, 1991).

The analysis involved the use of the inverse distance method (Mucha, 1991; Frąckiewicz et al., 2016). The application of that method required the preparation of a set of X, Y data representing the coordinates of the measurement points and Z—the elevation of the measured point and the value changes between successive sample measurements 1 and 2. The optimum interpolation network was selected iteratively.

The results of measurements 1 and 2 (Frąckiewicz et al., 2016) are represented by semivariograms (Fig. 1 and Fig. 2).

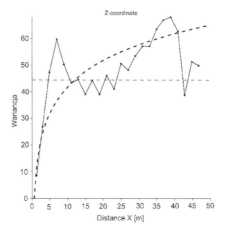

Figure 1. Semivariogram—measurement 1 (Frąckiewicz et al., 2016).

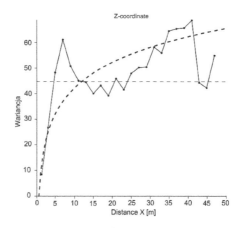

Figure 2. Semivariogram—measurement 2 (Frąckiewicz et al., 2016).

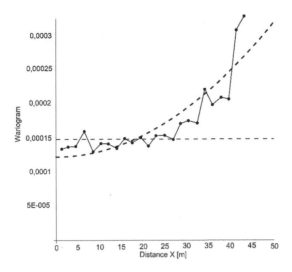

Figure 3.   Semivariogram of differences in the Z parameter between measurement 1 and measurement 2.

Semivariograms showing the variability of the landslide forms for measurements 1 and 2 are very similar, as was to be expected (the changes between the measurements are slight). The graphs of semivariograms for the case under consideration allow the conclusion that the share of the non-random component in shaping the landslide is dominant (96%), with the participation of the random component being at the level of 4%. The share of the non-random component decreases rapidly with the increasing distance between the measurement points. The semivariogram graphs reveal a certain periodic variability, which indicates that landslide forms may be repeatable.

Another semivariogram showed the differences in the Z parameter between measurement 1 and measurement 2 (Fig. 3) (Frąckiewicz et al., 2016).

The share of the random component is $U_L = 78\%$, while the share of the non-random component is $U_N = 22\%$. The semivariogram is of an "unlimited growth" type.

The obtained result is consistent with the expectations; the landslide process is predominantly of random nature. However, the non-random factor appears to quite a significant extent; thus, there is only little autocorrelation between particular areas where changes have been found. This autocorrelation can be explained by common causes of landslide movements, as well as by the possible effect of one-area displacements on activating the change process in neighboring areas. The share of the non-random component decreases with the increasing distance between the measurement points.

4   ASSESSING THE THREAT TO STABILITY AND DETERMINING THE REASONS FOR THE CHANGE

Based on the interpretation of the obtained graphical and statistical studies of the changes in the volume, slope gradient and cross sections, it was found that changes had definitely occurred. Their causes should be sought in numerous factors, often operating at the same time (Mucha, 1991).

The main cause of the slope instability is its geological structure. The measurement area is made up mainly of limestones, which are heavily cracked. Precipitations which cause the instability have a variable nature. Between the first control measurement and the primary measurement at the turn of June and July there were heavy, several-day rainfalls which could have caused a downflow of rock material. In the winter period between the first and second control measurements there occurred about 70 freeze-thaw cycles of the water contained in the rocks. The snowfall occurred in winter; the daytime temperature in that period remained slightly above 0°C, at night falling below 0°C. This can cause expansion of crevices and consequently mass movements. Weathering of rocks is a natural process deteriorating the properties of the examined form. Water and carbon dioxide facilitate the dissolution of rocks. Moisture, manifested in the form of numerous spots, facilitates the growth of vegetation and erosion of rocks. The vegetation in the form of grasses and small trees significantly interferes with the structure, expanding the existing crevices. The icefall formed every year in a part of the investigated section also has a destructive impact on the state of landslides. It is located in the area of the smallest changes, but freezing water has a destructive effect on the structure of the massif.

5 CONCLUSIONS

1. The use of scanning allows the assessment of landslide changes, the determination of the magnitude of movements and can be used for monitoring this type of forms. The resulting large set of points allows the use of comparative method "shape to shape" and partial elimination of sources of inaccuracy.
2. The usefulness of different statistical development methodologies for the comparative method "shape to shape" with the adopted resolution was evaluated. The studies confirmed the effectiveness of the method based on the measurements of the slope inclinations and the changes in the rock mass volume. The displacement values obtained using the profile method show both losses and gains in the rock mass between the particular measurements.
3. Semivariogram graphs for the analyzed sample measurements 1 and 2 permit the conclusion that the share of the non-random component in shaping the landslide is dominant. The share of non-random component decreases rapidly with increasing distance between the measurement points. The semivariogram graphs reveal a certain periodic variability, which indicates that landslide forms may be repeatable.
4. The semivariogram of the differences in the Z parameter between the measuring cycles has the nature of "unlimited growth" and indicates the predominant share of the random component. The obtained result is consistent with the expectations; the landslide process is predominantly of random nature. There is little autocorrelation between different areas showing the changes. This autocorrelation can be explained by common causes of landslide movements, as well as the potential effect of displacements in one area on activating the change process in neighboring areas.
5. The examined object demonstrates mass movements and requires further monitoring. Landslide monitoring should take place twice a year: in the Spring (March–April), after the snow cover disappearance, and in Autumn (September–October), after a period of precipitation.

REFERENCS

Duma P., Gajos K., Gil M., Godzisz B., Kaleta D., Krawczyk K., Salamon M., Siwiec K., Szewczyk J., Woś J., *Study of the condition of the selected landslides in the area of Kadzielnia,* in: Structure and Environment. Volume: 8 (2016), Journal: 1, pp: 64–74.

Frąckiewicz P., Oziewicz M., *Examination of a selected landslide in Kadzielnia in Kielce using scanning* [in Polish]. BSc dissertation under the supervision of I. Romanyshyn, Kielce University of Technology, Kielce 2016.

Mucha J., *Selected mathematical methods in mining geology* [in Polish]. AGH University of Science and Technology textbook nr 1215, Krakow 1991.

Advances and Trends in Geodesy, Cartography and Geoinformatics – Molčíková et al. (Eds)
© 2018 Taylor & Francis Group, London, ISBN 978-1-138-58489-1

Verification of horizontal circles quality of surveying instruments

V. Hašková

Department of Surveying, Faculty of Civil Engineering STU, Bratislava, Slovak Republic

ABSTRACT: Still higher demands on the quality and accuracy of geodetic works are required. In order to achieve the required precision of geodetic works in all areas it is necessary to use such instruments in the measuring process, whose characteristics are in good agreement with specified requirements. One of these features is the consistency, accuracy and repeatability of values of horizontal directions. The process for the quality verification of theodolite horizontal circles is based on the standard STN ISO 17123-3, some service centres or specialized laboratories perform the control of divided circles in accordance with the manufacturer requirements under the DIN and ISO standards or corresponding to the specific references of the laboratory. The automated etalon device EZB-3 of the Slovak Institute of Metrology in Bratislava allows the calibration of horizontal circles surveying instruments using a primary standard of planar angle of the Slovak Republic.

1   INTRODUCTION

In most of the tasks in geodetic practice we encounter with measurement of horizontal directions. Despite the progress in the development, manufacturing and modernization of electronic total stations the accuracy of horizontal directions measurement is influenced by errors, one of them is an error of horizontal circle dividing. Therefore, there is a requirement for testing and calibration of angle measuring instruments in appropriate period.

The test of these instruments si regulated by standard (STN ISO 17123-3). The accuracy of the angle measurement performed by the instrument is tested under field conditions in simplified or full test procedure with statistical tests. But in both methods we get a limited number of measurements. On the other hand, the tested geodetic instruments display during measurement a vast number of discrete values on their display unit and they must also be checked.

The calibration of angle measuring instruments requires comparison of an enormous number of angular values with the reference values implemented by the calibration device with a higher accuracy than the precision of a calibrated instrument. This procedure is not regulated by any standard in Europe at all. Such devices are usually operated by measurement equipment manufacturers or service centers and are not available for the wide public.

2   VERIFICATION AND CALIBRATION METHODS

There are several methods for exact angle determination. The system circular scale—microscope was widely used in the past. It requires a highly accurate circular scale and one or more microscopes for the scale readings (Bručas et al. 2006). Recently is being replaced by the method using digital rotary encoders as the reference measure. This method allows reducing the size of the test bench and good possibilities for its automation. It is a part of horizontal circle comparator at the ESRF in Grenoble, France (Martin 2010). Another method uses a precise tool, Moore's 1440 precision index, and consists of two serrated plates joined together to create the angle standard of measure. This method is characterized by high precision but is very difficult to automate (Bručas et al. 2006). Collimator system measurement is used in a compact laboratory test method recommended for all Leica Geosystems AG service centers. The combined measurement of horizontal and vertical angles is performed by means of five collimators suitably located in the space (Lovíšek 2002). The system polygon—autocollimator is widely applied in measuring technique and in geodesy instrumentation as well. The method is based on the precise multiangular prism—polygon, with 12 to 72 flat mirrors positioned at a very precise constant angle to each other, in connection with autocollimator for registration of measured rotation angle. The system with 18-edged polygon is a part of TPM-2 from Leica Geosystems AG and is used for the development of angular sensors, for the control of production precission and for verification of the theodolites angular measurement accuracy too (Lippuner et al. 2005). The system polygon—autocollimator is also used in the automated etalon device EZB-3 of the Slovak Institute of Metrology in Bratislava and is a part of a primary standard of planar angle of the Slovak republic.

3 EZB-3 STANDARD EQUIPMENT

The standard equipment EZB-3 is composed of OKT-315 Zeiss Jena rotation dividing table with servo-drive, where a calibrated polygonal prism is located, and two photoelectric autocollimators TA-80 Hilger & Watts. These elements are located on a cast iron base plate with anti-vibration deposition. The scheme of EZB-3 standard equipment is shown in Figure 1. Both autocollimators are in a base position pointed at the neighbouring function surfaces of the traverse and by its stepwise rotation the autocollimators measure the deviations of the normal positions to the traverse function surfaces regarding to the optical axes of autocollimators. The combined standard deviation of correction for polygonal prism calibration is $u_c = 0.06''$ (Mokroš 2001).

For calibration of horizontal circle is on the position of calibrated traverse located the 72-edged Starrett Webber polygonal prism with centric stored calibrated theodolite fitted with a plane mirror. The first autocollimator, which also serves as a part of rotation servo-driving of the dividing table, is focused on the polygonal prism. The second autocollimator is situated on the base plate with its optical axis at the height of the theodolite telescope axis (see Figure 2).

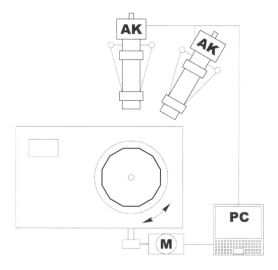

Figure 1. The EZB-3 standard equipment scheme.

Figure 2.   Calibration of Leica TCR407 on EZB-3.

4   MEASUREMENT AND DATA PROCESSING

The adjustment conditions of equipment EZB-3 described in (Mokroš 2001) must be fulfil before the measurement. Before starting the calibration measurement, it is necessary to select calibration cycle counts and calibration step. The calibration step should be an integral multiple of the nominal angle value between neighbour function areas of the polygonal prism. In the case of used polygonal prism is the nominal angle value between neighbour function areas 5°. One calibration step consists of both autocollimators data scanning into the measurement system PC and of the prism and theodolite turning of calibration step (multiple of 5°) with the deviation of 0.1″.

For the calibration was used the electronic total station Leica TCR407 with coded circular division for reading the horizontal angle. Our measurements were realised with 10° calibration step and 4 calibration cycle counts. For the measured data processing were created files with 10°, 20° and 30° calibration step and cycle counts from 1 to 4. Figure 3 shows the plot of horizontal circle corrections for Leica TCR407 calibration.

The aim of the measured data processing was to find a sufficient function type and its parameters values, which indicate the course of the horizontal circle correction values. We have focused to approximation based on polynomial and trigonometric function.

The polynomial function that describes correlation values of horizontal circle dependence from the reading variance could be written as follows:

$$k_i = \Theta_0 + \Theta_1.\alpha_i + \Theta_2.\alpha_i^2 + \ldots + \Theta_n.\alpha_i^n = \sum_{j=0}^{n} \Theta_j.\alpha_i^j \tag{1}$$

where $i = 1, 2, \ldots, m$; k = measurable correction; α = horizontal circle reading; and Θ_j = unknown polynomial parameters ($j = 0, 1, \ldots, n$).

The suitable polynomial degree is determined pursuant to various characteristics after unknown parameters estimation.

Trigonometric functions are used for approximation of periodical cases, i.e. cases with repetitive course at certain period. The case caused by one source could be described by simple sinusoid:

$$k_i = \Theta_0 + a_1.\sin\left(\frac{2\pi}{P}\cdot\alpha_i + \varphi_1\right) \tag{2}$$

where $i = 1, 2, \ldots, m$; Θ_0 = sinusoid drift in y axis line; P = period; a_1 = sinusoid amplitude; and φ_1 = period origin shift.

49

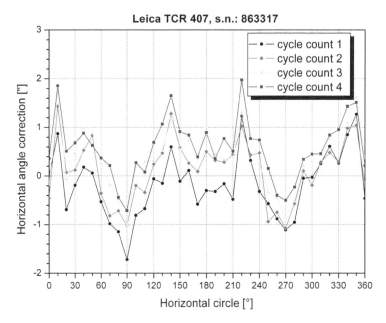

Figure 3. The plot of measured horizontal circle correction values for Leica TCR407.

5 RESULTS

The estimation of unknown parameters of polynomial and trigonometric series is solved by the least squares method (Hašková et al. 2010). Both functions were determined for each of correction values file which we got from the horizontal circles calibration results. The suitable degree of the polynomial function and the type of a trigonometric series was realized according to the value of coefficient of determination R^2 and adjusted coefficient of determination R^2_{adj} which indicate the rate of compliance of empirical measured data with the proposed models. The coefficient of determination value more closely to 1 means, that the proposed model represents better the empirical measured data. By addition of parameter which doesn't increase, the model quality the value of coefficient R^2 is non-decreasing, but the coefficient R^2_{adj} could decrease and get also negative value, because into the estimation of it comes the number of estimated parameters too.

The coefficients values R^2 and R^2_{adj} estimated as maximal for the number of calibration steps and cycle counts from calibration results of Leica TCR407 in model based on polynomial and trigonometric series are summarized in Table 1. Beside the polynomial series degree and value of trigonometric series multiple are in brackets given the statistical insignificant terms. They were eliminated from the adjustment results and rest parameters were estimated again.

From the results we can see that for Leica TCR407 the best reached result in the case of polynomial series is by 30° calibration step with 1 cycle count and the equation which defines the course of the horizontal circle correction values is in the form of polynomial function of 10th degree without parameter Θ_0:

$$\hat{k}_i = -0,823.\alpha_i + 6,811.10^{-2}.\alpha_i^2 - 2,188.10^{-3}.\alpha_i^3 + 3,646.10^{-5}.\alpha_i^4 -$$
$$-3,561.10^{-7}.\alpha_i^5 + 2,151.10^{-9}.\alpha_i^6 - 8,142.10^{-12}.\alpha_i^7 + 1,881.10^{-14}.\alpha_i^8 - \quad (3)$$
$$-2,426.10^{-17}.\alpha_i^9 + 1,337.10^{-20}.\alpha_i^{10}$$

Table 1.   Coefficients of determination for Leica TCR407.

Step/Cycle	Polynomial series			Trigonometric series		
	Degree	R^2	R^2_{adj}	Multiple	R^2	R^2_{adj}
10°/1	6 (0)	0.464	0.377	6 (6)	0.555	0.499
10°/2	6 (0)	0.490	0.408	6 (0,6)	0.629	0.595
10°/3	6 (0)	0.499	0.418	6 (0)	0.642	0.584
10°/4	6 (0)	0.493	0.412	6 (6)	0.631	0.585
20°/1	10 (0)	0.777	0.554	8 (0,6,8)	0.687	0.624
20°/2	10 (0)	0.811	0.622	6 (0)	0.827	0.761
20°/3	10 (0)	0.815	0.630	6	0.828	0.742
20°/4	10 (0)	0.803	0.606	6	0.817	0.725
30°/1	10 (0)	0.956	0.824	4	0.771	0.656
30°/2	10 (0)	0.950	0.800	4	0.849	0.773
30°/3	10 (0)	0.935	0.741	4 (0)	0.812	0.749
30°/4	10 (0)	0.945	0.780	4 (0)	0.829	0.771

Leica TCR 407, s.n.: 863317

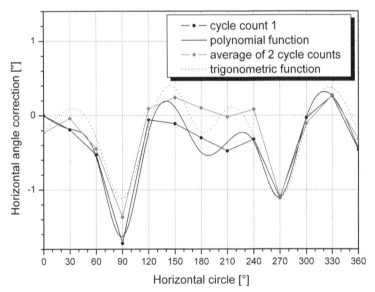

Figure 4.   Curves determining corrections course of horizontal circle for Leica TCR407.

In the case of trigonometric series is the best reached result by 30° calibration step with 2 cycle counts and the equation is in the form of trigonometric function with argument multiple 2 and 4:

$$\hat{k}_i = -0,22'' + 0,46'' \cdot \sin\left\{ 2 \cdot \left(\frac{2\pi}{P} \cdot \alpha_i \right) + 106,63° \right\} +$$
$$+ 0,45'' \cdot \sin\left\{ 4 \cdot \left(\frac{2\pi}{P} \cdot \alpha_i \right) + 266,00° \right\}$$

(4)

Graphic representation of curves determined by equations (3) and (4) define the course of the horizontal circle correction values for Leica TCR407 is in Figure 4.

6 CONCLUSION

The most used methods for verification and calibration of surveying instruments horizontal circles were discussed. The automated etalon device EZB-3 of the Slovak Institute of Metrology in Bratislava, which is a part of a primary standard of planar angle of the Slovak republic, has been described in more detail. This standard equipment was used for calibration of horizontal circle of electronic total station Leica TCR407. The calibration measurement data were ordered to files with 10°, 20° and 30° calibration step and cycle counts from 1 to 4. For calibrated total station the suitable function which determines the course of the horizontal circle correction values in dependence on the value of horizontal circle reading was estimated. The approximation function was based on polynomial and trigonometric function. According to the value of coefficients of determination the suitable degree of the polynomial function and the type of the trigonometric series for the electronic total station was selected. From the estimated results we can conclude that for the determination of the course of the horizontal circle correction values is sufficient the calibration of Leica TCR407 with 30° calibration step and up to 2 calibration cycle counts.

ACKNOWLEDGMENT

This work was financially supported by Cultural and Educational Grant Agency of the Ministry of Education, Science, Research and Sport of the Slovak Republic No. KEGA 037STU-4/2016.

REFERENCES

Bručas, D. & Giniotis, V. & Petroškevičius, P. 2006. Basic Construction of the Flat Angle Calibration Test Bench for Geodetic Instruments. *Geodesy and Cartography* 32(3): 5p.

Hašková, V. & Lipták, M. 2010. The Horizontal Circles Calibration Results Analysis. *Geodesy, cartography & geographic information systems 2010, Demänovská dolina, 07-09 September 2010:* 7p. Košice: F BERG TUKE.

Lippuner, H. & Scherrer, R. 2005. Die neue Theodolit-Prüfmaschine TPM-2 von Leica Geosystems. *Allgemeine Vermessungs-Nachrichten AVN* 5.

Lovíšek, I. 2002. Compact method for testing total stations. *Slovenský geodet a kartograf* 7(2): 30–35.

Martin, D. 2010. Instrument Calibration at the ESRF. *Proc. The XXIV FIG International Congress, Sydney, 11–16 April 2010.*

Mokroš, J. 2001. Calibration of theodolites horizontal circle. *Metrology in Geodesy*: 123–130. Bratislava: KGZA SvF STU.

STN ISO 17123–3: Optics and optical instruments—Field procedures for testing geodetic and surveying instruments—Part 3: Theodolites. 2010.

Advances and Trends in Geodesy, Cartography and Geoinformatics – Molčíková et al. (Eds)
© 2018 Taylor & Francis Group, London, ISBN 978-1-138-58489-1

Accuracy analysis of continual geodetic diagnostics of a railway line

J. Ižvoltová & T. Cesnek
University of Žilina, Žilina, Slovakia

ABSTRACT: Geodetic diagnostics of railway line belongs to the absolute diagnostics methods, which bring directional, longitudinal and height characteristics of an observed railway object. Modern automatized robotic techniques enable using continual measurements, which causes changing the methodology of data evaluation. The focus of the paper is directed on the analysis of accuracy of continual deformation observations of railway line, which comprises a comparison of numerical methods intent on 2D and 3D modelling of an observed railway line.

1 INTRODUCTION

Diagnostics of an absolute and relative geometrical track position distinguishes two types of data acquisition: discrete and continual. The discrete diagnostics is based on an observation of a single-track parameter and the other ones are used to define in post-processing. The absolute position of the observed object is given in the local or global coordination system realized by the railway benchmarks (Ižvolt, et al. 2015). The technology of continual observations of a track construction is based on continual data acquisition, which brings "clouds" of data assume to be manipulate by the numerical methods. The main difference of the discrete and continual diagnostics consists in the fact that the discrete one requires uniquely determination of the particular observed point in each observed period and the continual one anticipates the mutual comparisons of the huge of data. This new approach of observation brings the new view on an estimation process by using adequate numerical methods to ensure the specified accuracy and to define the real track position or changes. The focus of the paper is in accuracy analysis of mathematical methods applied to estimate the spatial track position by using clouds of data derived from the continual monitoring of ballast-less track construction. The aim of research described on the paper is in considering the accuracy and consequently the reliability of 2- and 3-dimensional numerical methods, based on principles of analytical geometry and regression analysis.

2 ABSOLUTE DIAGNOSTICS METHOD

The above mentioned mathematical methods was implied on the terrestrial geodetic observations acquired by robotic total station TRIMBLE S8. It is video-assisted robotic total station utilizing Trimble VISION technology, which means that it sees everything without a trip back to the instrument and select targets with just a tap of the controller screen. So, measurements are drawn to the video image and surveyor can be certain to never mess a shot he needs. The total station involves also Trimble FineLock technology to detect targets without interference from surrounding prism and Trimble SurePoint accuracy assurance to correct instrument pointing. The a-priori precision of **TRIMBLE S8** is involved in its technical parameters, where producer defines angle accuracy about 0,3 miligons, distance measurement accuracy in standard prism mode is about 1 mm + 2 ppm and distance measurement accuracy in tracking mode is about 4 mm + 2 ppm. Geodetic observations were realized on 800 meters long ballast-less track using the standard prism mode. The track position was determined relative to railway benchmarks defined in the national coordinate system.

The railway line diagnostics was realized in semi-annual observation cycles (periods), whereby the first measurement was performed just after railway building acceptance. The main aim of the geodetic observations was to find spatial track changes in particular observation periods, which are related to the basic observation realized in the first period. While the directional changes are defined by the difference values of track axis in longitudinal and transversal direction, the height changes represent the differences of no surmounted track part (Šíma et al. 2007). Because of the specific technology of continual observations, the longitudinal track changes are impossible to capture. So, only the transversal and height track changes was evaluated by using principles of analytic geometry, regression analysis and similarity transformation to compare their results and reached accuracy.

3   METHODS OF ANALYTIC GEOMETRY

Analytic geometry models geometric objects as points, straight lines, circles, planes, surfaces, etc. and defines the relations between them by using linear and nonlinear formulations. The principles of methods of analytic geometry are based on manipulation with the geometric object by means of kartesian coordinates of the particular points defined in two or three dimensions (Villim et al. 2013). For defining the transversal and height track changes, we used linear formulations of points, lines and planes and their perpendicular distances. However, the two and three-dimensional techniques of analytic geometry were applied to compare the final track shifts and their accuracy.

3.1   *Application of 2-dimensional model*

The track transversal difference between two observation periods is defined by the shortest distance between a line in the plane given by the general equation

$$ax + by + c = 0, \tag{1}$$

with a, b, c represented unknown line parameters and a point $P_0(x_0, y_0)$, which represents the position of a periodic observation:

$$d_{2D} = \frac{|ax_0 + by_0 + c|}{\sqrt{a^2 + b^2}}. \tag{2}$$

The real constants a, b, are coordinates of a normal vector of a line $\mathbf{n} = (a, b)$ and a, b cannot be zero both unless c is also zero. If $a = 0$ and $b \neq 0$, the line is horizontal. Values x_0, y_0 are coordinates of a point observed in the particular observation period and x, y both of them are adjusted coordinates comes from Least Square Estimation used in Mixed Gauss-Markov model defined by stochastic formulas:

$$f\big(\mathbf{E}(l), \mathbf{x}\big) = 0, \ \mathbf{D}(l) = \sigma^2 \mathbf{Q}, \tag{3a, b}$$

where $\mathbf{E}(l)$ represents the theoretical expectation of the measured values l and $\mathbf{x}$ is unknown vector. In case of parametric equations of a line

$$x = x_0 + at, \ y = y_0 + bt, \tag{4}$$

where $t \in \mathfrak{R}$, the distance between the point $P_0(x_0, y_0)$ and the line passing through $P(x, y)$ in direction of normal vector $\mathbf{n}$ is defined by the formula

$$d_{2D} = \sqrt{\frac{\left| \begin{matrix} x_0 - x & y_0 - y \\ a & b \end{matrix} \right|^2}{a^2 + b^2}}. \tag{5}$$

54

Because of 2-dimensional model, height shifts of the observed track are estimated separately as the difference

$$\Delta z = z_0 - z_Q, \tag{6}$$

where z_0 is the third coordinate of the point P_0 and z_Q is defined as height of the closest point Q on the original line to the point P_0 estimated by a linear interpolation method. The other coordinates x_Q, y_Q of the point Q are calculated according to the following equations:

$$x_Q = \frac{b(bx_0 - ay_0) - ac}{a^2 + b^2}, \quad y_Q = \frac{a(-bx_0 + ay_0) - bc}{a^2 + b^2}. \tag{7}$$

3.2 Application of 3-dimensional model

Spatial data acquisition assumes spatial evaluation, which starts with 3D model construction in every differential part of the observed railway line. Analytic geometry provides similar formulas for the perpendicular distance from the point $P_0(x_0, y_0, z_0)$ to the line defined in three dimensions by the parametric equations (4) augmented of z-axis:

$$\frac{x - x_0}{a} = \frac{y - y_0}{b} = \frac{z - z_0}{c}. \tag{8}$$

The spatial distance between the point $P_0(x_0, y_0, z_0)$ and the line passing through $P(x, y, z)$ in direction of normal vector $\mathbf{n} = (a, b, c)$ is as follows:

$$d_{3D} = \sqrt{\frac{\left| \begin{matrix} \frac{x_0 - x}{a} & \frac{y_0 - y}{b} \end{matrix} \right|^2 + \left| \begin{matrix} \frac{y_0 - y}{b} & \frac{z_0 - z}{c} \end{matrix} \right|^2 + \left| \begin{matrix} \frac{z_0 - z}{c} & \frac{x_0 - x}{a} \end{matrix} \right|^2}{a^2 + b^2 + c^2}}. \tag{9}$$

In case of real constants $a \neq 0$, $b \neq 0$, $c \neq 0$, the spatial distance between point and original plane defined by its general formula

$$ax + by + cz + d = 0 \tag{10}$$

is estimated by the equation:

$$d_{3D} = \frac{|ax_0 + by_0 + cz_0 + d|}{\sqrt{a^2 + b^2 + c^2}}. \tag{11}$$

4 ESTIMATION OF ACCURACY OF 2D AND 3D MODEL

A-priori accuracy represents standard deviation σ_0, which involves observation errors depended on used instrumentation, methodology, atmospheric conditions, benchmarks precision, etc. Due to the observed data manipulation and processing, we obtain a-posteriori accuracy, which cumulates all measured and processing errors in accordance to principles of Law of Propagation (Zavada et al. 2014). The processing errors are involved in covariance matrices estimated by Least Square Method applied in Mixed Gauss-Markov model (3a, b), of which deterministic formula is

$$\begin{pmatrix} \mathbf{k} \\ -\mathbf{dx} \end{pmatrix} = \begin{pmatrix} \mathbf{B}^T \mathbf{Q} \mathbf{B} & \mathbf{A} \\ \mathbf{A}^T & 0 \end{pmatrix}^{-1} \begin{pmatrix} \mathbf{u} \\ 0 \end{pmatrix} = \begin{pmatrix} \mathbf{Q}_{kk} & \mathbf{Q}_{xk} \\ \mathbf{Q}_{xk}^T & \mathbf{Q}_{xx} \end{pmatrix} \begin{pmatrix} \mathbf{u} \\ 0 \end{pmatrix}, \tag{12}$$

where k is $(r \times 1)$ Lagrange vector, **A** and **B** are known matrices, u is $(r \times 1)$ known vector, **dx** is $(k \times 1)$ vector of unknown parameters and **Q** is weighted matrix of measured values. Covariance matrix of unknown parameters, which represents parameters of 2D and 3D model of railway line is estimated according to the formula

$$\mathbf{D}(x) = -\sigma^2 \mathbf{Q}_{xx} \tag{13}$$

and covariance matrix of adjusted values is estimated from the following equation:

$$\mathbf{D}(\bar{l}) = \sigma^2 \mathbf{Q}_{ll}, \tag{14}$$

where $\mathbf{Q}_{xx}$ and $\mathbf{Q}_{ll}$ are cofactor matrices estimated as follows:

$$\mathbf{Q}_{xx} = -\left(\mathbf{A}^T (\mathbf{B}^T \mathbf{Q} \mathbf{B})^{-1} \mathbf{A} \right)^{-1}, \mathbf{Q}_{ll} = \mathbf{Q} - \mathbf{Q} \mathbf{B} (\mathbf{B}^T \mathbf{Q} \mathbf{B})^{-1} \mathbf{B}^T \mathbf{Q}. \tag{15a, b}$$

If redundancy is $n - k$, the unit variance can be estimated according to the formula:

$$\sigma^2 = \frac{\mathbf{v}^T \mathbf{P} \mathbf{v}}{n - k}. \tag{16}$$

5   RESULTS

The above mentioned methodology of processing the transversal and height track shifts was applied on the differential parts of 800 meters long ballast-less track observed by the continual geodetic technique. A differential part of the observed track was about 10 meters long with 26 particular points. Mixed Gauss-Markov model was constructed from the slope-intercept formula of a straight line and plane. This simplification caused only formal changes of formulas (2), (5), (9) and (11) to precise estimate distance between $P_0(x_0, y_0)$ and straight line:

$$d_{2D} = \frac{|kx_0 + q - y_0|}{\sqrt{1 + k^2}} \tag{17}$$

and $P_0(x_0, y_0, z_0)$ and plane

$$d_{3D} = \frac{|kx_0 + qy_0 + w - z_0|}{\sqrt{1 + k^2 + q^2}}, \tag{18}$$

which represent the transversal track shifts between first and periodic observations.

For the illustration, there are results from one differential part of the observed locality in the Fig. 1, which involves the track shifts estimated from 2D and 3D model. Detailed results analysis indicates comparable values of transversal shifts of the observed part of railway line, which differences are mostly within the precision of used methodology defined by the value:

$$\sigma_{max} = 2\sigma_d. \tag{19}$$

Standard deviation σ_d means the a-posteriori accuracy of track shift determination. The particular parameters of both of 2D and 3D models and their accuracy characteristics are in Table 1.

Height differences of the observed railway line are displayed in Fig. 2. The numerical values are displayed in Table 1. For the comparison both of methods of analytical geometry, we involve also similarity transformation method into estimation process, which results are

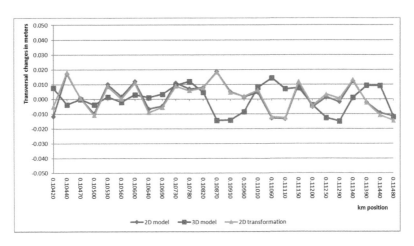

Figure 1.   Transversal track changes in meters.

Table 1.   Transversal and heights changes of track part determined by using 2D model 3D model and 2D transformation method.

km position	Transversal track changes in meters			2D-3D in mm	Heights track changes in meters			2D-3D in mm
	2D model	3D model	2D transf.		2D model	3D model	2D transf.	
0.10420	−0.0117	0.0076	−0.0053	*−19.3*	−0.0059	−0.0135	−0.0060	*7.6*
0.10440	0.0173	−0.0037	0.0180	*21.0*	0.0030	0.0067	0.0030	*−3.7*
0.10470	0.0005	−0.0002	0.0006	*0.7*	0.0010	0.0012	0.0010	*−0.2*
0.10500	−0.0097	−0.0038	−0.0108	*−5.9*	0.0060	0.0098	0.0059	*−3.8*
0.10530	0.0101	0.0014	0.0087	*8.7*	−0.0001	−0.0015	0.0000	*1.4*
0.10560	0.0016	−0.0020	0.0001	*3.7*	0.0020	0.0040	0.0020	*−2.0*
0.10600	0.0120	0.0029	0.0109	*9.1*	−0.0050	−0.0079	−0.0050	*2.9*
0.10640	−0.0066	0.0011	−0.0090	*−7.7*	−0.0020	−0.0031	−0.0020	*1.1*
0.10690	−0.0047	0.0034	−0.0056	*−8.1*	−0.0040	−0.0073	−0.0039	*3.4*
0.10730	0.0110	0.0093	0.0088	*1.7*	−0.0100	−0.0193	−0.0100	*9.3*
0.10780	0.0070	0.0121	0.0057	*−5.1*	−0.0150	−0.0270	−0.0150	*12.1*
0.10820	0.0079	0.0045	0.0077	*3.3*	−0.0050	−0.0096	−0.0050	*4.5*
0.10870	0.0187	−0.0143	0.0183	*33.0*	0.0139	0.0282	0.0141	*−14.3*
0.10910	0.0050	−0.0141	0.0045	*19.1*	0.0138	0.0279	0.0140	*−14.1*
0.10960	0.0014	−0.0085	0.0018	*9.9*	0.0070	0.0155	0.0071	*−8.5*
0.11010	0.0045	0.0080	0.0061	*−3.4*	−0.0100	−0.0180	−0.0099	*8.0*
0.11060	−0.0129	0.0144	−0.0120	*−27.3*	−0.0130	−0.0274	−0.0130	*14.4*
0.11110	−0.0133	0.0070	−0.0126	*−20.3*	−0.0041	−0.0111	−0.0040	*7.0*
0.11150	0.0104	0.0078	0.0119	*2.7*	−0.0080	−0.0158	−0.0080	*7.8*
0.11200	−0.0053	−0.0040	−0.0046	*−1.3*	0.0050	0.0090	0.0050	*−4.0*
0.11250	0.0013	−0.0126	0.0031	*13.9*	0.0110	0.0236	0.0110	*−12.6*
0.11290	−0.0019	−0.0150	0.0003	*13.1*	0.0120	0.0269	0.0120	*−15.0*
0.11340	0.0123	0.0011	0.0130	*11.3*	−0.0020	−0.0031	−0.0020	*1.1*
0.11390	−0.0023	0.0093	−0.0024	*−11.6*	−0.0071	−0.0163	−0.0070	*9.3*
0.11440	−0.0094	0.0092	−0.0108	*−18.5*	−0.0080	−0.0172	−0.0081	*9.2*
0.11480	−0.0115	−0.0121	−0.0145	*0.6*	0.0140	0.0261	0.0140	*−12.1*

represented by the third curve in the Figs. 1 and 2. The method of similarity transformation with the controlled scale factor is used to rotate the coordination system into railway direction to determine the pure transversal deformations without longitudinal influence.

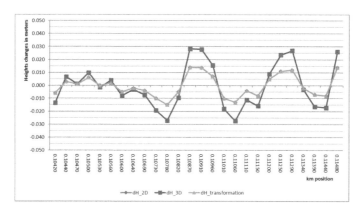

Figure 2.   Track heights changes in meters.

6   CONCLUSIONS

The main task of the paper was accuracy analysis of the diagnostics of continual observations of railway line, which comprises analysis of two methods of analytical geometry. The described methods are based on the construction 2D and 3D models to compare the transversal and height differences (shifts) between the first (basic) measurement and the periodic one. Final analysis of results indicates on the fact that while the differences between 2D model and 2D transformation represents only the random noise the differences between 2D and 3D application are significant, however the heights changes seems to be closer. The final accuracy of the estimated parameters derived from the 2D model is represented by the standard deviation $\sigma = 0,3$ mm. The accuracy of unknown parameters in the used 3D model is given by the values $\sigma = 44$–77 mm. Such an inconvenient differences can explain by using unsteady matrix of used parameters caused probably by the small dispersion of z-axis coordinates. For this reason, it seems to be more precise to use separately positional and height estimation method to reach the more reliable results. However, the track changes are not considerable. Its transversal and heights changes are in the frame of confidence interval, which results by using statistical hypothesis testing providing normal distribution.

ACKNOWLEDGEMENTS

This article is the result of the implementation of the project VEGA 1/0275/17 "Application of numerical methods to define the changes of geometrical track position", supported by the Scientific Grant Agency of the Ministry of Education, science, research and sport of the Slovak Republic and the Slovak Academy of Sciences.

   This article is the result of the implementation of the project ITMS 26220220156 "Broker centre of air transport for transfer of technology and knowledge into transport and transport infrastructure", supported by the Research & Development Operational Programme funded by the ERDF.

REFERENCES

Ižvolt, L., Hodas, S., Šestáková J.: Railway buildings 1. Design, building, construction of railway lines and stations. Handbook, EDIS 2015, ISBN 978-80-554-1122-4, 561 p.
Šíma, J., Koťka, V., Pisca, P., Seidlová, A.: Geodetic work for reconstruction and building of narrow-gauge railway. In: XIII. International Slovak-Poland-Russian geodetic days, Liptovský Ján 2007: ISBN 978-80-969692-0-3, p. 84–88.
Villim, A., Mužík, J.: Analysis of satellite and terrestrial measurements in 3D geodetic network for transport infrastructure, Juniorstav 2013, 15, Brno 2013.
Zavada, F., Stankova, H., Cernota, P., Lucan, L., Havlicova, M.: Statistical data analysis used by measurement testing. International Multidisciplinary Scientific GeoConference Surveying Geology and Mining Ecology Management, SGEM 2(2), 2014, pp. 459–471.

Advances and Trends in Geodesy, Cartography and Geoinformatics – Molčíková et al. (Eds)
© *2018 Taylor & Francis Group, London, ISBN 978-1-138-58489-1*

Verification of the quality of selected electro-optical rangefinders according to STN ISO 17123-4: 2013

J. Ježko & Š. Sokol
Department of Surveying, FCE STU Bratislava, Slovakia

ABSTRACT: The article describes the results of the verification procedures and quality of selected Electro-Optical Rangefinders (EOR). The results of the verification of electro-optical rangefinders that are part of the Universal Measuring Stations (UMS) Leica TS30 and Trimble S8 are presented. The contribution also deals with the possibilities of using the geodetic (longitude) baselines of some European states for the verification and control of EOR. The main part of the contribution is devoted to the principle and procedure of EOR control according to the standard STN ISO 171230-4: 2013. Presented are methods of verification and calculation of the estimated parameters of the 1st and 2nd order for these rangefinders. This part is followed by the implementation of test measurements of selected EOR.

1 INTRODUCTION

The quality verification and the accuracy comparison of the surveying instruments can be done using different procedures. One of the options to verification is the international standards usage (ISO 17123), which specify the testing procedures of the determining and precision estimating of the surveying instruments and other equipments for measurements. The output is verification, which declare the usage suitability of individual instruments for their purpose and their comparison with uniform process. Another option is usage of geodetic baselines in terrain.

2 QUALITY AND SURVEYING INSTRUMENTS VERIFICATION

Before the surveying, it is important to know if the used instruments precision meets the requirements of the task. Verification and testing measurements are used to determine this. The individual procedures and conditions for verification measurements are defined in the International Technical Standards of the ISO 171 Council, or it is possible to verify the quality of the rangefinder in terrain at the baseline (Ježko, J. 2008).

2.1 Some geodetic baselines in the terrain and their usage for verification and calibration of rangefinders

One of the main characteristic of electro-optical rangefinders (EOR) is the accuracy. The parameters reported by manufacturer are generaly obtained by multiple measurements in the laboratory conditions. Geodetic terrain baselines were built in previous periods and they allow the independent quality verification of the EOR precision.

Koštice—the Czech Republic national standard of length
The baseline consists of 12 pillars with forced centring. The baseline configuration has 66 measurable reference lengths in the range of 25 to 1450 m (Ježko, J. 2016).

Gödöllö—national comparison baseline in Hungary
The size of the geodetic baseline in Gödöllö was determined using the Väisälä interferometer. The baseline consists of 7 pillars, lengths are in the range of 6 to 864 m. 5 pillars have underground marks (Figure 1). The accuracy of the individual sections varies from ± 0,03 mm to ± 0,10 mm (Ježko, J., Sokol, Š. & Bajtala, M. 2007).

Figure 1. The stabilization of the length geodetic baseline Gödöllö.

Nummela—the length calibration baseline in Finland
The baseline is similar to Hungary baseline. The baseline configuration is realised by 5 lengths (24 m, 72 m, 216 m, 432 m, 864 m). The reference accuracy of the baseline is about 0.07 mm for the longest distance (Ježko, J., 2016).

The physical technical institute in Germany baseline
The baseline was built by the Physical Technical Institute in Germany. Its length is 600 m. It is unique, because of its technical equipment, which consists of 60 temperature sensors, 6 humidity sensors and 2 air pressure sensors (Ježko, J. 2016).

Hlohovec geodetic comparison baseline (SR)
On the Slovakia territory was built a geodetic comparison baseline at Hlohovec. The baseline consists of 5 pillars (Z1 – Z5). There were 10 combinations to lengths measurements. The measurement consisted of a measurement combination from each pillar (Ježko, J., Sokol, Š. & Bajtala, M., 2007).

3   VERIFICATION OF THE QUALITY OF EOR ACCORDING TO STN ISO 17123-4:2013

3.1   *General standard requirements*

The test results are influenced by atmospheric conditions. It is necessary to measure meteorological information for determining the atmospheric corrections to the distance measurements. Standard STN ISO 17123 describes two procedures. During the tests it is necessary to choose the approach, which is closer to the expected requirements of the particular measurement (Ježko, J., 2008, STN ISO 17 123-4:2013).

3.2   *Simplified testing approach*

A simplified testing approach is based on a limiting number of measurements. If it is necessary to have more accurate estimation of the EOR accuracy, it is recommended to execute more accurate method (test), i.e. complete testing approach. The testing approach requires a testing field with verified distances. If this testing field is not available, it is necessary to determine the unknown distances with more accurate EOR than tested device. If the more accurate device is not available, it is necessary to do complete testing approach (STN ISO 17 123-4:2013).

3.3   *Complete testing approach*

Complete testing approach is based on the distance measurements in all combinations on the test field without nominal values. The standard deviation of the distance measurement is determined by least square method (LSM) in all distance combinations.
   The procedure can also be used to determination of (STN ISO 17 123-4:20137):

- The accuracy of usage of the EOR by one measuring team with only one device and its attachment at the time,
- The accuracy of usage single device uses in a wider time range,
- The accuracy of usage of each of several EOR devices to compare their accuracy under the same conditions.

Statistical tests should be used to determine whether the experimental standard deviation "s" belongs to the group of theoretical standard deviation "σ", and whether two tested samples belong to the same group and whether is the value of the addition constant (correction to zero point) "δ" is zero or is equal to predetermined value δ_0. The test baseline has a length from 300 m to 600 m and consists of seven stabilized points in a straight line (configuration of the baseline shows Figure 2).

3.3.1 *Compilation of the testing baseline of 21 different lengths*
A suitable configuration of the testing baseline can be achieved, if six distances $d_1 - d_6$ are formed by a process in which the measurements of all 21 combinations of lengths have different values according to formula (1):

$$d_1 = \frac{d}{63}, d_2 = 2d_1, d_3 = 4d_1, d_4 = 8d_1, d_5 = 16d_1, d_6 = 32d_1. \tag{1}$$

3.3.2 *EDM testing baseline*
If the value on the EDM addition constant is influenced by a systematic error, the distances of the baseline points should be selected using the scale (EDM wavelength) (Ježko, J. 2016, STN ISO 17 123-4:2013). All distances between the seven points (Figure 3) should be measured in one day, using depended centring method, with good visibility and without cloudy.

The recommended configuration will be achieved, when six distances between seven baseline points will be compute using following formula:

$$\beta_0 = \frac{d - 6,5.\lambda}{15}, \tag{2}$$

where d is the total length of the projected baseline, λ is the wavelength of the EDM (derived from the modulation frequency EDM), $\lambda/2$ is a unit of length (measuring unit) of EDM and

$$\beta = \mu.\frac{\lambda}{2}. \tag{3}$$

The integer value of μ is selected. The value of number β is so close as possible to β_0, with

$$\gamma = \frac{\lambda}{72}. \tag{4}$$

Six lengths of the baseline and the total length is compute by:

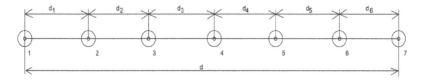

Figure 2. Baseline configuration of the complete testing approach.

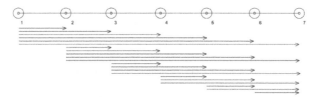

Figure 3. Measured distanced in the baseline.

61

Figure 4. Baseline configuration according to complete testing approach.

$$d_1 = \lambda + \beta + 3.\gamma,$$
$$d_2 = \lambda + 3.\beta + 7.\gamma,$$
$$d_3 = \lambda + 5.\beta + 11.\gamma, \quad \quad (5)$$
$$d_4 = \lambda + 4.\beta + 9.\gamma,$$
$$d_5 = \lambda + 2.\beta + 5.\gamma,$$
$$d_6 = \lambda + \gamma,$$
$$d = 6.\lambda + 15.\beta + 36.\gamma.$$

3.3.3 *Experimental testing of selected EDM according to STN ISO 17123-4:2013*

Test field, which complied the parameters of standards STN ISO 17 123-4:2013 has been implemented near the Faculty of Civil Engineering (Figure 4). The baseline had a total length about 300 m (Ježko, J. 2016, STN ISO 17 123-4:2013). The oblique lengths were measured from the total station station according to observation plan (points 1–6). Each length was measured four times: two times in the first position of the telescope and two times on the second position of the telescope.

These values were calculated to the horizontal lengths. Another mathematical reduction was that the horizontal lengths were reduced by the value of the turn from the vertical plane, which was created from baseline start point and baseline end point. The reduced average value of the horizontal lengths from four measurements entered to the processing. 21 combinations of the 7 baseline points consist the measurement. The atmosphere parameters (air temperature, atmospheric pressure, humidity) were recorded on the device's station.

4 THE RESULTS PROCESSING ACCORDING TO STN ISO 17 123-4:2013

The measurements, $\bar{x}p,q$ (not corrected lengths = readings from EDM) have been corrected by the atmospheric corrections. These corrected length values $x_{p,q}$ joint the alignment by the LSM. Unknown parameters are six lengths $y_{1,2}$, $y_{2,3}$, $y_{3,4}$, $y_{4,5}$, $y_{5,6}$, $y_{6,7}$ and addition constant δ (STN ISO 17 123-4:2013). 21 observational equations are formulated as follows:

$$x_{1,2} + r_{1,2} = 1.y_{1,2} + 0.y_{2,3} + \cdots + 0.y_{6,7} - 1.\delta$$
$$x_{1,3} + r_{1,3} = 1.y_{1,2} + 1.y_{2,3} + \cdots + 0.y_{6,7} - 1.\delta \quad \quad (6)$$
$$\vdots$$
$$x_{6,7} + r_{6,7} = 0.y_{1,2} + 0.y_{2,3} + \cdots + 1.y_{6,7} - 1.\delta$$

In the matrix description of the linear system:

$$x + r = F(y), \text{residues} : r = A.y - x, \quad \quad (7)$$

x is the observables vector with dimensions 21×1, y is unknown estimated parameters vector with dimensions 7×1, r is residual vector with dimensions 21×1,

$$x = 1 = \begin{pmatrix} x_{1,2} \\ x_{1,3} \\ \vdots \\ x_{5,7} \\ x_{6,7} \end{pmatrix} \quad y = \begin{pmatrix} y_{1,2} \\ y_{2,3} \\ y_{3,4} \\ y_{4,5} \\ y_{5,6} \\ y_{6,7} \\ \delta \end{pmatrix} \quad r = \begin{pmatrix} r_{1,2} \\ r_{1,3} \\ \vdots \\ r_{5,7} \\ r_{6,7} \end{pmatrix}. \tag{8}$$

To solve of the unknown estimated parameters is used Gauss-Markov model, which satisfied the LSM condition. It is defined by:

$$y = \left(\mathbf{A}^T.\mathbf{P}.\mathbf{A} \right)^{-1} \mathbf{A}^T.\mathbf{P}.\mathbf{x}, \tag{9}$$

if we consider that all measurements have the same weight and that they are uncorrelated, the weights matrix can be defined as a matrix with ones. The shape of normalized matrix than is:

$$\mathbf{N} = \mathbf{A}^T.\mathbf{A}. \tag{10}$$

The final shape of the estimated parameters:

$$y = \mathbf{N}^{-1}.\mathbf{A}^T.\mathbf{x}. \tag{11}$$

According to this, the standard deviation of one measured length s_0 is defined by:

$$s_0 = \sqrt{\frac{\mathbf{r}^T.\mathbf{r}}{v}}, \text{ where:} \tag{12}$$

$d = 21$ is number of observed lengths, u is the number of the unknown estimated parameters and $v = d - u = 14$ is the number of the degrees of freedom. The experimental standard deviations of each estimated length $\mathbf{y_k}$ and the estimation of the additional constant δ are calculated by multiplying of the diagonal elements of the cofactor matrix $\mathbf{Q}$ by standard deviation of one measurement.

$$\mathbf{Q} = \mathbf{N}^{-1} \tag{13}$$

$$s(\mathbf{y_k}) = s_0 \times \sqrt{\mathbf{Q_{k,k}}}, k = 1,...,6 \tag{14}$$

$$s_\delta = s_0 \times \sqrt{\mathbf{Q}_{7,7}}. \tag{15}$$

The standard uncertainties of all lengths (estimated parameters):

$$\mathbf{u(y_k)} = \mathbf{s(y_k)}, k = 1,...,6 \tag{16}$$

and the standard deviation of the addition constant δ:

$$u_\delta = s_\delta \tag{17}$$

$$u_{\text{ISO-EOD}} = s_0. \tag{18}$$

Table 1. The estimated values of the 1st and 2nd order of Trimble S8.

	d	σ_d	δ	σ_δ	s_0
	[m]	[mm]	[mm]	[mm]	[mm]
d_1	4.7620	0.3	−36.4	0.2	0.5
d_2	9.5239	0.3			
d_3	19.0528	0.3			
d_4	38.0972	0.3			
d_5	76.1920	0.3			
d_6	152.3758	0.3			

Table 2. The estimated values of the 1st and 2nd order of Leica TS 30.

	d	σ_d	δ	σ_δ	s_0
	[m]	[mm]	[mm]	[mm]	[mm]
d_1	4.7617	0.2	−0.2	0.2	0.4
d_2	9.5245	0.2			
d_3	19.0522	0.2			
d_4	38.0965	0.2			
d_5	76.1911	0.2			
d_6	152.3725	0.2			

Table 3. Tested values of selected EDM.

UMS (EDM)	s_0 (length uncertainty) [mm]	δ (Addition Constant − AC) [mm]	σ_δ (AC uncertainty) [mm]	EDM accuracy according to manufacturer	Addition constant
Trimble S8	0,5	−36,8	0,2	1 mm + 1 ppm	−36,4 mm
Leica TS30	0,4	−0,2	0,2	0,6 mm + 1 ppm	0,0 mm

The results of the estimated parameters with the characteristics of the estimated parameters accuracy are shown in Table 1 and Table 2. (Ježko, J., 2016, STN ISO 17 123-4:2013).

5 CONCLUSION

The paper presented a part of the results of testing according to standards STN ISO 17 123–4 using complete testing method with Trimble S8 and Leica TS 30. According to the achieved results of the quality control, we can say, that the devices, which were controlled by complete testing method complied with the requirements of the standards, i.e. achieved control results (measurements uncertainty) corresponds with manufacturer data, including the addition constant determination of reflection system (GPH1). Summary and comparison is in Table 3 (Ježko, J., 2016).

REFERENCES

Ježko, J. 2008. *Testovanie a kalibrácia geodetických prístrojov z pohľadu technických noriem. In.: Interdisciplinárne aplikácie geodézie, inžinierskej geodézie a fotogrametrie.* Bratislava, KG SvF STU, 2008, 10 s., (CD ROM) ISBN 978-80-227-2938-3.
Ježko, J. & Sokol, Š. & Bajtala, M. 2007. *Kalibrácia elektronických diaľkomerov na dĺžkovej porovnávacej základnici v teréne.* Acta Montanistica Slovaca: roč.2007, číslo 3, str.393–396.
Ježko, J. 2016. Porovnanie elektrooptických diaľkomerov podľa STN ISO 17 123-4. In *Geodézie a Důlní měřictví 2016 [elektronický zdroj]: XXIII. konference Spoločnosti důlních měřičů a geologů., Karolinka, ČR, 19.–21. 10. 2016.* 1.vyd. Ostrava: VŠB-TU Ostrava, 2016, USB kľúč, 13 s. ISBN 978-80-248-3977-6.
STN ISO 17 123-4:2013, *Optika a optické prístroje/Postupy na skúšanie geodetických prístrojov, Časť 4:Elektrooptické diaľkomery.*

Advances and Trends in Geodesy, Cartography and Geoinformatics – Molčíková et al. (Eds)
© 2018 Taylor & Francis Group, London, ISBN 978-1-138-58489-1

Landslide movement monitoring as an element of landscape protection

K. Krawczyk & J. Szewczyk

Geomatic and Energy Engineering, Faculty of Environmental, Kielce University of Technology, Kielce, Poland

ABSTRACT: Landslide movements, both of natural and technogenic or antropegenic origins are a growing global problem. Monitoring of these movements can contribute to the development of systems warning against their intensification. Surveying methods are important solutions applied in this type of monitoring. Advances in monitoring technologies have led to the development of comprehensive techniques of measuring the state of landslides. These measurements provide input for the development of surveying data bases which are a fundamental source in analyzing the behavior of areas threatened by landslides. The paper presents the research results concerning the area of Kadzielnia in Kielce. The site protected because of its geological values, paleontological qualities and landscape attributes bears the marks of changes caused by landslides. The paper stresses the complementary nature of the results obtained using various surveying techniques and to the possible use of the data for the protection against the emerging threats.

1 INTRODUCTION

Landslides and threats they pose are becoming a significant global problem. In the built-up areas the phenomenon under consideration causes a real threat to human life and health. It can also severely damage technical and transportation infrastructure. Moreover, landslides degrade the area, preventing further use of the ground for crops or land development. Monitoring hazardous areas followed by taking preventive measures help to reduce the potential damage caused by the phenomenon.

One of the areas with intensive landslide movements is Kadzielnia—a strict reserve of inanimate nature located in the center of Kielce. The monitored area of stricte reserve is 0.6 ha, the area (partially monitoring) of all object – 13.5 ha, the highest elevation reaches 295 m.a.s.l., the height difference is up to 53 m. Today's nature reserve is situated in the former site of Devonian limestone mine operating from the 17th century until 1962. After the intensive limestone extraction the eastern slope was the only part that remained from the original hill, together with the remnants of the south-west slope with the adjacent mound (presently called Wzgórze Harcerskie – Scouts Hill) and Skałka Geologów (Geologists Rock), separated by a deep quarry pit.

Although several years have passed since the quarry ceased its operation, the post-extraction condition still poses threat both for the reserve landscape and for visitors. The condition of the slopes that were subjected to quarrying is far from stability. Thus, the activation of landslide processes—commonly rapid ones—is possible. The activity of landslides is indicated by the accumulation of masses of soil and rock debris at the foot of the quarry pit walls. Hence, it is necessary to monitor the changes occurring on the slopes in order to determine the hazards related to the activation of the landslide process. Surveying methods are most common in landslide monitoring.

In 2014 the first monitoring of selected landslides was conducted in Kadzielnia within the cooperation between the Faculty of Environmental, Geomatic and Energy Engineering of the Kielce University of Technology and Kielce Geopark. To date, 23 monitoring cycles of the selected landslides have been performed, including 18 using a scanning total station, 4 using a laser scanner and 1 using a drone taking photogrammetric images. The performed monitoring cycles permitted the identification of the areas potentially threatened with the development of the landslide processes. The observations resulted in ten BSc dissertations, publications in English (with BSc students as co-authors) (Duma et al. 2016, Klimczyk et al. 2016, Kowalczyk et al. 2016), and two cumulative reports (handed to the Kielce Geopark authorities).

2 THE NOTION OF LANDSLIDE AND THE REASONS FOR MASS MOVEMENTS

It is widely accepted that the landslide is a displacement of rock masses, formed in the conditions of constant contact with the ground, mainly due to the action of gravity along the slide surface. Disrupting the balance of the two forces—the one maintaining the slope (related to the strength of the form) and the other pressing the slope—cause the occurrence of landslides and other destructive mass movements. The loss of slope stability is due to numerous factors. If stability is taken into account, the hydrogeological conditions and the geological structure, mainly related to the layer arrangement, are particularly dangerous causes of the landslide forming process.

The landslides examined by the authors (with the total of nine distinguished landslide areas) to be found within the Kadzielnia reserve are of the insequent type. The shearing surface partly belongs to the weathered rock area and partly runs along the fissure surface. Colluvium consists of detritus-block material (granular soils, weathered rock, rock boulders). Due to the almost vertical slope wall, the mass movement in the area of Kadzielnia can also be classified as falling.

3 SURVEYING METHODS OF LANDSLIDE MOVEMENT MONITORING

The assessment of landslide movements, their magnitude, direction and velocity requires quantitative information based on periodic (at regular intervals) measurements. The objective is to determine the range of the landslide, the velocity and direction of its movement. The monitoring methods are divided into two main groups: the surface ones regarding the movements on the landslide and the subsurface ones.

Surface methods for testing landslides are connected with defining the movements of the surface-representing points, which occurred on the landslide between successive measurements. As far as the methods of the landslide surface representation are concerned, one can distinguish two types of surveys: the point surveys and the area surveys. The first type consists in the monitoring of measurement points permanently stabilized on the surface whereas the area surveys refer to the whole area or to unstabilized points. Point surveys permit determining the movements of selected points with a high level of accuracy. Area surveys are slightly less accurate but they allow for the monitoring of arbitrary parts or the whole landslide area (Maciaszek et al. 2015).

Point surveys include:

– Angular-linear surveys consisting in the measurement of horizontal coordinates by a polygon method (traverses tied to points of a higher order network) or by polar method (from selected control points)
– Situational and altitude surveys consist in the measurements of the polar coordinates (horizontal angle, horizontal distance) from each stabilized point and elevation (Δh—trigonometric levelling) to an arbitrary point signaled by means of a mirror. These methods are applied when the masses move fast and there is a possibility of damaging the research point.

Area surveys include the following:

– Tacheometry;
– Scanning tacheometry;
– Photogrammetry;
– Laser scanning;
– LIDAR (Light Detection and Ranging), which is airborne laser scanning;
– Measurements using satellite technology (GNSS systems);
– Satellite radar interferometry – InSAR;
– Combined methods GPS and InSAR;
– Terrestrial Interferometry;
– The hybrid method combines elements of both the point methods and the area ones.

Subsurface methods for measuring landslides are complementary for surface methods and are used mainly for large landslides. The necessity to use subsurface methods is due to the differences between the slope surface deformations and the subsurface deformations. The factors influencing these differences include the type of movement and shape as well as the position and the number of slide surfaces, the degree and nature of the landslide disintegration, the nature and geotechnical characteristics of soils and rocks and the degree of the ground waterlogging. The thorough landslide monitoring should be based on both the surface and subsurface methods.

4 CHOICE OF THE LANDSLIDE MONITORING METHOD AND THE PERFORMED MEASUREMENTS

The following methods (with the availability of measurement equipment being the main factor determining the choice) were selected to conduct the research on landslides in Kadzielnia:

– satellite techniques – static measurement (to determine the coordinates of the network points);
– laser scanning of the landslide surface using a scanning total station;
– laser scanning of the landslide surface by means of a laser scanner;
– taking photogrammetric images with the use of a drone.

The reference tacheometric survey was performed in May–June 2014, followed by further surveys in October–November 2014, June–July and October–November 2015 and November 2016. Laser scanning was performed in September 2014 and in November 2016, whereas the drone aerial photographs was taken in October 2016. Within a few months between measurement cycles the temperature of the atmospheric air was high, which caused the intense heating of the rocks, as well as frequent and heavy rainfalls; winter periods saw the formation of icefall (artificial), having an impact on the rock coherence. Such atmospheric conditions increase the intensity of physical weathering, whose effect is leaching and washing away of rock material, chipping rocks, etc.

Measurement works related to the study of the landslides included project of a network, stabilization of control points; measurements of network and measurement of the landslide area by means of the listed methods.

The stabilized geodetic control in the form of two columns (for the forced instrument centering) and two points in the form of pins has been measured several times using linear and angular method and RTK as well as geometric leveling; position errors of points after alignment amounted to 3 to 9 mm, elevation errors – up to 2 cm. The survey involved the use of the Quick Station 1 A electronic total station; the satellite observation – the use of the Sokkia GRX1receiver.

The TOPCON scanning total station Quick Station 1 A was used to conduct the tacheometric survey of the landslide area, with the automatic mode applied in the majority of observations. The observed points were not stabilized. The monitoring was mostly carried out in regular squares (with sides of about 0.50 m) or rectangles (with sides measuring 1.0 × 1.5 m), or in profiles distant from each other by 2 to 4.5 m. The number of points for which the coordinates were determined ranged from 600 to 2200 in different landslides. The average error in determining the location of pickets was about 3 cm, therefore 0.10 m was adopted as the MPE (crucial to recognize the change significance on the landslide surface).

Three landslides were subject to laser scanning (in September 2014 and November 2016). The OPTECH scanner ILRIS 3D was applied in the measurements in 2014 and STONEX X300 was used in 2016. ILRIS 3D scanner was applied within the cooperation with the company Czerski Trade Polska S.A. Further observations will be performed using the STONEX scanner, along with the JRC 3D Reconstructor software to develop the scanning results. The number of points in the cloud ranged from 10 to 21 million.

Photogrammetric images of one landslide were taken by the mini drone Phantom 3 Professional. It is an unmanned aircraft of the new generation of quadrocopters. Its signal strength varies depending on weather conditions. The embedded camera is capable of recording videos in 4 K resolution at 30 frames per second and take pictures of 12 megapixels using an array of 1/2.3" CMOS. The flight data are automatically recorded in the embedded aircraft memory and include telemetry, aircraft status information, and other parameters.

5 DETERMINATION OF THE DISPLACEMENT MAGNITUDE OF THE LANDSLIDE AREAS

The determination of the displacement magnitude in the landslide area was based on the comparison of two surfaces (measured in the reference cycle and then in the next measurement cycle) within the specified profiles using the "shape to shape" method. When interpreting the results, it was adopted that the value of the MPE (m_g) in determining the location of the point was 0.10 m as the value of the displacement significant for the accuracy of monitoring.

The monitoring results were analyzed using profiles, with the number of profiles for individual landslides ranging from 10 to 14, and the distance between adjacent profiles from 2 m to 4.5 m. The profiles were developed in the AutoCAD Civil. This stage included also the development of landslide models for primary and secondary measurement, which additionally allowed demonstrating the differences between the models and the determination of the mass movement areas (sample model shown in Fig. 1).

The profile charts provided the basis for distinguishing the regions with changes on the surface of landslides. Fig. 2 shows an example of such an area marked accordingly.

The results of the observations performed by different methods were compared with each other. A high level of compatibility and also complementary character of the information obtained from scans and photogrammetric observations is clearly noticeable. Scanning by

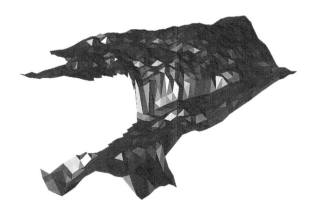

Figure 1. The Eastern slope of Geologists Rock—pseudo-dimensional model of the object created using AutoCAD Civil 3D (Duma et al., 2015).

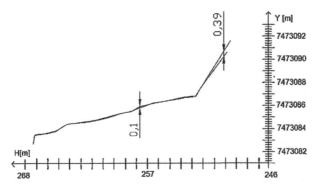

Figure 2.   Eastern landslide. Sample comparison of changes along the profile 14 (Duma et al., 2015). (Y-coordinates in the Polish 2000 system, H – heights in the Kronstadt-86 system).

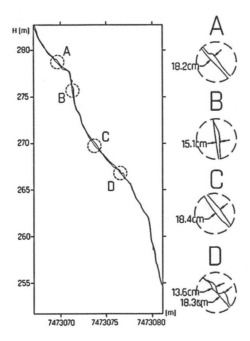

Figure 3.   Comparison of laser scanning survey results (solid line) and the results obtained from photogrammetric images taken by the drone (broken line) in the representative landslide profile (horizontal axis: Y-coordinates in the Polish 2000 system, H – heights in the Kronstadt-86 system).

means of a scanning total station provides a general insight into the landslide area and enables highlighting the regions where surface changes have occurred. Scanning using a laser scanner completes the picture, allowing the location of detailed changes and more precise determination of the character of their distribution.

A similar role should be performed by photogrammetric images. Their comparison with the scanning results (Fig. 3) reveals that the maximum difference does not exceed 20 cm, with differences in values greater than 5 cm occurring only locally, in rockfall areas and in places where no laser scanning survey points were available. The data indicate the suitability of these comprehensive measurements to determine the displacement magnitude and a clear consistency of landslide images obtained by all the methods.

6 CONCLUSIONS

1. The landslides in Kadzielnia, which are a consequence of many years of aggregate quarrying, may pose a threat to visitors and adversely affect the landscape qualities of the nature reserve. Determining the possibility of the emergence of a such a threat requires research, including monitoring by surveying methods.
2. Surface methods were applied to perform the observations. Observations were conducted from four points of the established control, permanently stabilized, allowing convenient measurement (including two points with forced instrument centering). The measurements were performed using the scanning Topcon total station QS1 A. Due to the almost vertical slope there was no possibility of permanent stabilization of the measurement points. Reflectorless measurement was made directly to the surface of the examined area, yielding satisfactory results in terms of accuracy.
3. Experimental observations with ILRIS 3D and STONEX X300 laser scanners were also carried out. Scanning enables measurements with higher precision than those conducted by conventional methods; it also allows the measurement of objects in hard-accessible places. The scanning result represents both a surface model and its image.
4. Taking into account the errors of geodetic control points and errors in determining the location of the pickets, an MPE, amounting to 0.10 m., was assumed. The displacements of landslides exceeding this value should be recognized as significant, if not related to the individual points (which can be interpreted as a lack of coverage of the pickets in successive measurement cycles), but associated with a set of several neighbouring points.
5. The monitoring results were analyzed using the profiles method, with the number of profiles for each landslide ranging from 10 to 14, and the distance between adjacent profiles from 2 m to 4.5 m. The profiles were developed in AutoCAD Civil. The landslide models for the primary and secondary measurements were also developed, which permitted demonstrating differences between models and determining the location of mass movements. The Excel programme was applied to compare displacements between particular pickets.
6. The comparison reveals that the results obtained by laser scanning are consistent with those obtained by the development of the images taken by the drone. The maximum difference does not exceed 20 cm, with differences in values greater than 5 cm occurring only locally, in rockfall areas and in places where no laser scanning survey points were available.
7. The results indicate the suitability of these comprehensive measurements to determine the displacement magnitude and a clear consistency of landslide images obtained by all the methods.

REFERENCES

Duma P., Gajos K., Gil M., Godzisz B., Kaleta D., Krawczyk K., Salamon M., Siwiec K., Szewczyk J., Woś J., *Study of the condition of the selected landslides in the area of Kadzielnia*, in: Structure and Environment. Volume: 8 (2016), Journal: 1, pp: 64–74.

Klimczyk P., Krawczyk K., Ptak J., Sawadro K., Siarek A., Szewczyk J., Walkiewicz K., Stępień A., *Determining the shape and volume of the post-mining basin in Kadzielnia area*, in: Structure and Environment. Volume: 8 (2016), Journal: 2, pp: 55–64.

Kowalczyk L., Krawczyk K., Makuch A., Mosiołek J., Piśkiewicz B., Pluta M., Proboszcz A., Szewczyk J., Świdzicka K., Wirecka A., *Study of the condition of the selected underground caves in Kadzielnia area*, in: Structure and Environment. Volume: 8 (2016), Journal: 3, pp: 197–205.

Maciaszek J., Gawałkiewicz R., Szafarczyk A., *Surveying methods of landslides investigation* [in Polish], AGH, Krakow 2015.

Advances and Trends in Geodesy, Cartography and Geoinformatics – Molčíková et al. (Eds)
© 2018 Taylor & Francis Group, London, ISBN 978-1-138-58489-1

Measuring and creating of the mapping documentation of the part of the Josef Gallery

T. Křemen

Department of Special Geodesy, Faculty of Civil Engineering, Czech Technical University in Prague, Czech Republic

ABSTRACT: Each mine work in the Czech Republic must have its own mine surveying documentation. Since the Josef Gallery now does not have the main mining surveyor and the original documentation of mine operation is incomplete and outdated, the Department of Special Geodesy has undertaken to create its new mapping documentation for the needs of the mine. This paper describes measuring of the accessible parts of the Čelina – east area and creation of their mine mapping documentation. The peculiarity of this part is a huge cavern, which passes through all floors of mine and which is not drawn on any map. The measurement was carried out by the laser scanning method. As a result of the mapping of the accessible part of the Čelina – east area, Josef Gallery is a 3D model in the point cloud form, digital map documentation in 3D and 2D and a printed basic mining plan.

1 INTRODUCTION

The Josef Gallery is a mine that was excavated within the geological exploration of gold-bearing deposits during the 1980s and that revealed one of the largest gold deposits in our country. Besides the short experimental extraction the deposit was not extracted due to the negative impact on the surrounding nature and the environment (immediate closeness to the Slap Dam). The gallery was closed at the beginning of the nineties of 20th century. It was reopened in 2007, when the Centre of Experimental Geotechnics (CEG) launched a research and training center offering the opportunity to solve experiments and teachings in a real underground gallery. The Centre of Experimental Geotechnics is a research and pedagogical department of the Faculty of Civil Engineering of the Czech Technical University in Prague (CTU in Prague). The Josef Underground Facility and the Josef Underground Research Center (Josef URC) run in the Josef Gallery. The Gallery is lent for this activity by the owner of the mine which is Ministry of the Environment of the Czech Republic.

The current position of Josef Gallery for The State Mining Administration is different from conventional mine. The gallery is led as an abandoned mine work. There is no extra mining in the pit, only the safety work needed to access and operate the mine. The gallery has its mine manager (závodní), but there is no need to occupy many other positions needed to run a mine. These include the main mining surveyor. It is a reason why the gallery has not actual mapping documentation. The department of Special Geodesy of Faculty of Civil Engineering of CTU in Prague helps solve this problem. The department organizes here teaching of students in the field of geodesy and cartography (Jiřikovský, 2016), (Urban, 2016), (Urban, 2015) and carries out some basic surveying work for CEG. It ensures the rebuilding of the surveying net (Braun, 2012) and new mapping of the whole of gallery. The mapping carries out according accessing of particular parts of gallery and their importance. The opening part of the main corridor, Čelina – west and part of the Mokrsko – west region were mapped already by classical methods.

The accessible part of the Čelina – east area is most recent opened part of the mine. This area was measured and its map documentation was created. The work was realized within

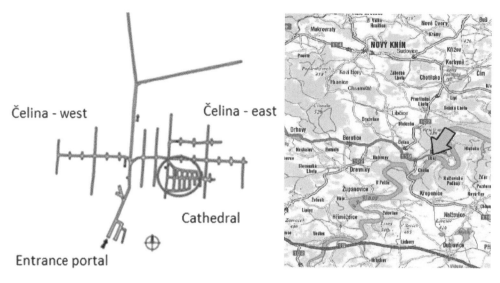

Figure 1.   Location of Cathedral and location of Josef Gallery on the map.

two diploma theses of Filip Špaček and Zbyněk Pražák. 3D laser scanning was used for the work (Štroner, 2013).

2   LOCALITY

The Josef Gallery is located 50 km south of Prague between the Čelina and Smilovice villages near Příbram near the Slapy dam. It is part of the golden district Psí Hory. The mine consists of the regions Čelina and Mokrsko, which are further divided into the western and eastern parts. These two areas are connected by 2 km long main corridor. Above main corridor there are two more levels in the Čelina – east area. The total length of the corridors is 7853 m. The height of the overlay is between 90 m and 150 m.

The mapped area is located in the accessible part of the Čelina – east. This is the most complicated part of the whole of Josef Gallery passing through all three levels. There is a large cavern called Underground Cathedral, whose dimensions can be said to be impressive. Its length is over 30 m, it is about 8 m wide and its height exceeds 40 m. The cathedral was the main target of mapping, because of unknown reasons there is no available mine map. Proof can be a segment of the area plan of Čelina, in which the location of the cathedral is marked (Fig. 1).

In addition to the cathedral, the main entrance corridor of Čelina – east with adjoining crossings between the main corridor of the mine and the cathedral at a 0 m level, the nearest corridors around the cathedral at a +20 m level and the main corridor with crossings, which leads to the surface by secondary portal, at a + 40 m level were measured.

3   MEASUREMENT

Surveying net was built before mapping. The points of the net were stabilized in the ceiling or in the floor. Six new points were stabilized to the walls in the cathedral. The surveying net was measured by a spatial traverse that was extended in some areas into a simple spatial net (Štroner, 2014). Heights were still determined by precise levelling. The net was adjusted using robust methods (Hampacher, 2015). The surveying net was used for georeferenc- ing of the mapping into the S-JTSK coordinate system (Souřadnicový systém Jednotné

Figure 2.   FARO Focus 3D X 130 (left), Trimble TX8 (right).

trigonometrické sítě katastrální; Datum of Uniform Trigonometric Cadastral Network) and the Bpv datum (Výškový systém baltský – po vyrovnání; Baltic Vertical Datum – After Adjustment).

3D laser scanning method was used for measurement of the area. Two 3D scanning systems FARO Focus 3D X 130 and Trimble TX8 were used. Both systems were lent by GEOTRONICS Prague company. Principle of measurement of both scanners is the spatial polar method. FARO system has phase electronic distance meter. It measures distances up to 120 m (manufacturer specification, real range of measurement is up to 60 m). The field of view is 300° in vertical direction and 360° in horizontal direction. The scanning speed is up to 976000 points per second. The scanner is equipped with a two-axis. The scanner weight is 5 kg. Trimble TX8 system has pulse electronic distance meter. It measures distances up to 120 m in standard mode and up to 340 m in extended range mode. Standard deviation of the distance measurement is 2 mm. The field of view is 317° in vertical direction and 360° in horizontal direction. The scanning speed is up to one million points per second. The scanner is equipped with a two-axis. The scanner weight is 11 kg (Fig. 2).

Measurement was carried out during one day. All corridors were scanned by FARO system and cathedral was scanned by TX8 system. Forty scans were measured in all corridors. Quality of measurement 4 (it is technical parameter) and density of scanning 12 mm in distance 10 m were set for the FARO system. The speed of scanning was 122 thousand points per second and time of scanning was 3 minutes. Travers was measured in the corridors by set for the three-tripod system. Ending points of the three-tripod system were signalized by spheres with diameter 200 mm. Between each two tripods there were signalized two points by sphere with diameter 145 mm to ensure correct registration. Three scans were measured in the cathedral. The range 120 m and density of scanning 5.7 mm in distance 30 m were set for system TX8. The speed of scanning was one million points per second and time of scanning was 10 minutes. Six chessboard targets were used in the cathedral.

4 PROCESSING

Registration of the scanned point clouds was carried out in the Cyclone software. It was necessary to model the scanned spherical targets and to number correctly. Then the point clouds were registered into the S-JTSK coordinate system and the Bpv datum. Mean absolute error of the registration was 2 mm. Correct connection of the scans was checked by horizontal and vertical cuts through individual corridors and the cathedral. There were used 6 planar black and white targets and 111 spherical targets for the registration. Figure 3 shows an overview of the final point cloud of the Čelina – east area.

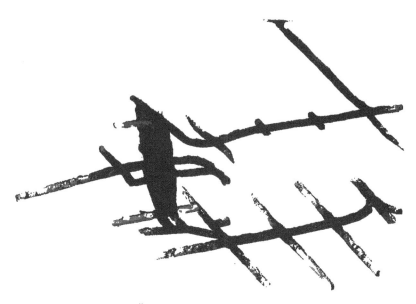

Figure 3.   Final point cloud of the Čelina – east area.

Mine mapping documentation was created according the Decree No. 435/1992 (vyhláška 435/1992 Sb., 1997). It was produced in the form of a basic mine plan in the digital form, which contains some layers of thematic mine maps. Map was created in the DULMAP. It is extension of the software Microstation. The DULMAP supports the drawing of map marks and lines according to Decree No. 435/1992 and according to the rules for creation of large scale maps (ČSN 013411). Graphical data are stored in 3D drawing files. The data for mine mapping documentation in this work were taken by the laser scanning method and therefore a spatial line drawing from the cloud point in the Cyclone software was created to draw maps in the DULMAP. According to the Decree, the basic mine plan (BMP) is prepared for each floor (horizon). Three maps were created, for the basic floor 0 m, for the floor +20 m and for the floor +40 m. Scale 1: 500 was taken from the existing parts of the BMM, which were completed in the last two years in Josef Gallery. The horizontal cut was chosen at a height of 1 m above floor. The cross sections were selected at the intersection of the corridors in all direction, in places with significant change of the profile shape and in each 5 m in the regular corridors. Others components were drawn into specialized layers which are needed for drawing of the map such as lights, switches, rails, etc.

The final drawing of the map was done in the DULMAP. Two Drawings, a Mine Map, and Engineering Networks map were created for each map of the floor, in which individual elements were plotted according to the categories defined by the line type and the map marks according to the Decree. An example of the processing in DULMAP is shown in Figure 4. Map sheets (MS) were generated, the map layout of which corresponds to division in S-JTSK on a given scale. These new drawings were still in 3D.

5   RESULTS

The result is 3D Drawings of the Basic Mining Plan for each floor containing layers of thematic mine maps. Maps were divided by map sheets. Each floor is shown on two map sheets. For the printed outputs, the map frame was shifted so that the whole area of the floor was on one map sheet. Segment of the map is shown in Figure 5.

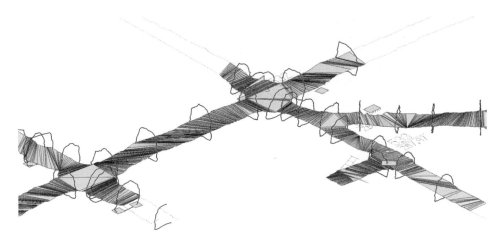

Figure 4.   An example of the processing in DULMAP.

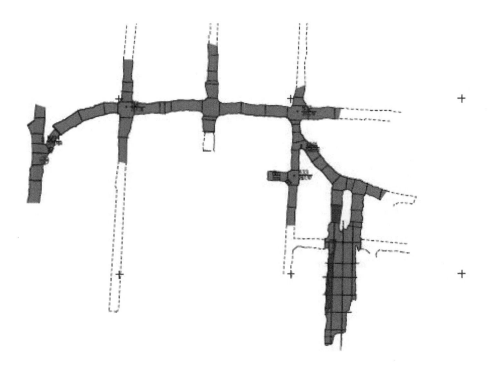

Figure 5.   Segment of the printed map of the basic floor 0 m.

6   CONCLUSION

The aim of the project was new mapping of the accessible parts of the Čelina – east area in the Josef Gallery. The measurement was carried out by two 3D laser scanners, Trimble TX8 an FARO Focus 3D X 130. The advantage of the map processing from laser scanning was more detailed drawing and possibility of choosing the cross section position during the processing. Another advantage of laser scanning was the measurement of spatially compli-cated and less accessible areas such as cathedral in the mapped area. The laser scanning time of measurement was significantly shorter than time of conventional methods. Although the processing of data from laser scanning was more time consuming than the processing of classical measurements, overall, the process of measuring and processing the Basic Mining Plan was more efficient, faster and, above all, with a much higher quality output. As a result of the mapping of the accessible part of the Čelina – east region, Josef Gallery is a 3D model in the form of a point cloud, digital map documentation in 3D and 2D and a printed Basic Mining Plan.

ACKNOWLEDGEMENT

This work was supported by the Grant Agency of the Czech Technical University in Prague, grant No. SGS17/067/OHK1/1T/11 "Optimization of acquisition and processing of 3D data for purpose of engineering surveying, geodesy in underground spaces and laser scanning".

REFERENCES

Braun, J. Štroner, M. & Třasák, P. 2012. Experimentální určení přesnosti záměry při nivelaci. *Geodetický a kartografický obzor* vol. 58/100, iss. 10, p. 226–236. ISSN 0016-7096.

Hampacher, M. & Štroner, M. 2015 *Zpracování a analýza měření v inženýrské geodézii*. Praha: CTU Publishing House, p. 336 ISBN 978-80-01-05843-5.

Jiřikovský, T. 2016 Testování laserového provažovače Foif JC100 v podmínkách štoly Josef In: *Sborník referátů Mezinárodní konference Geodézie a Důlní měřictví 2016 – XXIII. konference SDMG*. Ostrava: Společnost důlních měřičů a geologů.

Štroner, M. & Třasák, P. 2014 P. A New Procedure For Exact Joint Adjustment Of The Three Dimensional Network In The Large Areas. In: *14th International Multidisciplinary Scientific Geoconference SGEM 2014 – Informatics, Geoinformatics and Remote Sensing—Conference Proceedings Volume II—Geodesy & Mine Surveying*. Sofia: STEF92 Technology Ltd., pp. 3–10 ISSN 1314-2704 ISBN 978-619-7105-11-7.

Štroner, M., Pospíšil, J., Koska, B., Křemen, T., Urban, R., Smítka, V. & Třasák, P. *3D skenovací systémy*. Praha: CTU Publishing House, p. 396, ISBN 978-80-01-05371-3.

Urban, R. & Jiřikovský, T. 2016. TUNNELING MEASUREMENT IN UEF JOSEF USING TRIMBLE S8 In: *16th International Multidisciplinary Scientific Geoconference SGEM 2016 Book 2 Informatics, Geoinformatics, and Remote Sensing Volume II*. Sofia: STEF92 Technology Ltd., pp. 689–696. ISSN 1314-2704. ISBN 978-619-7105-69-8.

Urban, R. & Jiřikovský, T. 2015 ACCURACY ANALYSIS OF TUNNELING MEASUREMENTS IN UEF JOSEF In: *15th International Multidisciplinary Scientific GeoConference SGEM 2015*. Sofia: STEF92 Technology Ltd., pp. 35–42. ISSN 1314-2704. ISBN 978-619-7105-35-3.

Vyhláška č. 435/1992 Sb. *Českého báňského úřadu o důlně měřické dokumentaci při hornické činnosti a některých činnostech prováděných hornickým způsobem ve znění vyhlášky Českého báňského úřadu č. 158/1997 Sb." úplné komentované znění*, 1997. Ostrava: Montanex. ISBN 80-85780-88-7.

Advances and Trends in Geodesy, Cartography and Geoinformatics – Molčíková et al. (Eds)
© 2018 Taylor & Francis Group, London, ISBN 978-1-138-58489-1

Influence of image compression on image and reference point accuracy in photogrammetric measurement

M. Marčiš & M. Fraštia
Faculty of Civil Engineering, Slovak University of Technology, Bratislava, Slovakia

ABSTRACT: One of many factors influencing the accuracy of photogrammetric outputs is the quality of image data. This is especially important in high precision measurements, where we meet with subpixel measurement accuracy and therefore the circular or coded targets are used. The accuracy of Least Square Matching (LSM) method depends on similarity between template and search image and target diameter and can reach theoretical value of 0.005 pixel. Similarity of images is strictly given by perspective, projective and optical distortions and radiometric differences. We know that image compression changes radiometric properties of some pixels which could have an impact on measurement of image coordinates by the LSM method. In this article we empirically determine the influence of various image compression on the measurement accuracy of points in the image plane and on resulting 3D coordinates in reference system. The influence of different size and shape of circle targets in combination with image compression is also tested. The coded targets on a special testing frame were evaluated using the multi-image convergent photogrammetry. The RAW images were converted into various image formats with different compression and the influence of this compression on point measurement accuracy was estimated.

1 INTRODUCTION

Algorithms based on JPEG compression are the most commonly used in the practical photography. Especially, in the fields of close-range and UAV photogrammetry, significant HDD space savings can be reached while large image blocks are acquired. The impact of image compression on the accuracy of photogrammetric measurement was described in various works theoretically and practically. JPEG compression ratios in the order of 1:10 (Jaakola & Orava, 1994) are basically recommended, however subjective differences in the compressed images are discernible at compression rates in the order of 1:15.

The discrete cosine transform JPEG (baseline) compression standard is still the most used despite the effort of Joint Photographic Expert Committee to supersede it by the JPEG 2000 with Wavelet-based compression method. Detailed experiments with image quality measurement based on the PSNR (peak-signal-to-noise-ratio) showed that Wavelet compression influences the compressed images less than JPEG baseline compression. Yet the JPEG baseline compression ratio of 1:5 should not negatively influence the photogrammetric work (Kiefner & Hahn, 2000).

However, relative sufficient requirements of image quality for DTM generation may be deficient for high accurate applications (Akcay et al., 2017). In order to achieve maximum accuracy circular or coded targets are used and sub-pixel accuracy of image coordinates is expected (Šedina & Pavelka, 2016), (Urban et al., 2015). Various methods of targets center computation can be used (e.g. LSM, Star or Zhou operators) delivering different results. The size and the flattened shape of circle targets should also not be neglected (Luhmann et al., 2013).

The aim of this contribution is to enhance the knowledge of impact of the image quality on photogrammetric accuracy. We empirically determined the influence of different image compressions of most common image formats (TIFF, JPEG, JPEG 2000) on the accuracy of measurement of coded targets in image plane. Also, impact on 3D reference coordinates was analyzed considering the size and the shape of the originally circle targets.

The images used in this research were taken by a 36 megapixel Nikon D800E DSLR camera equipped with a Nikkor 35 mm AF-S ED 1:1.8G lenses. The images were saved originally in Nikon raw NEF-format and then converted to the tested image formats. All tested variations were processed in the same way using the same settings in PhotoModeler software created by the Eos Systems Inc.

2 EXPERIMENT

2.1 Acquiring data

A special testing frame with 100 coded targets was created (Figure 1. left) in laboratory conditions. Another 8 coded targets (8 mm in diameter) around the frame were used to provide the reference coordinate system.

40 targets were printed in different sizes (3, 5, 7, 9, 11, 13, 15 and 17 mm in diameter) and placed in the center of the frame surface approximately parallel to the image plane of the first image. Assuming distance of the camera to object, mentioned targets were projected on the image plane in the sizes of 7, 12, 17, 22, 26, 31, 36 and 41 pixels.

60 targets (8 mm in diameter) were positioned on the frame using hand-made paperboard segments so the circle targets could be projected on the image plane under various angles in one shot and therefore they were projected to the image as an ellipsis. Value of this elliptical distortion of targets expressed by the ratio of ellipse axis a/b could be also evaluated (Figure 1. right).

11 images were taken under the conditions required for the processing of multi-image convergent photogrammetry (Figure 2). A heavy tripod and time shutter release were used to minimize the influence of camera shaking on the image sharpness. The aperture of f10 provided sufficient depth of field for the entire 3D scene. The scale for reference coordinate system was defined by measuring the lengths between 8 reference coded targets around the frame using a tape.

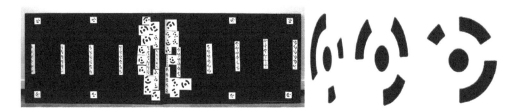

Figure 1. Testing frame with 108 coded targets (left) and examples of targets shape changes due to different convergent angle—different elliptical b/a ratio (from left to right: 0.3, 0.5 and 0.9).

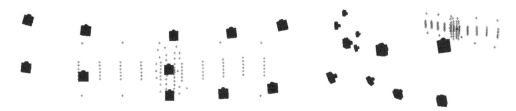

Figure 2. Camera positions relatively to the testing frame—front view on the left, perspective view on the right.

2.2 *Image processing*

In first step the original raw images were converted to the format TIFF 16-bit (lossless) – this set of images served as the reference set to which all other solutions were compared. Tested variations with different quality factor (QF) were as follows: TIFF 8-bit (lossless), TIFF 8-bit (LZW), JPEG baseline (100% - 1% QF) and JPEG 2000 (lossless, 100% - 1% QF).

Photogrammetric processing was done in the PhotoModeler software after image conversion. Detection of automatic coded targets was performed using the LSM method. The fit error threshold value was set to 0.2, to reduce the number of rejected points. Software Photo-Modeler uses as default setting the green channel to detect the targets to minimize the chromatic aberration effect—so this setting was kept. Next, image and 3D reference coordinates of all points were exported and analyzed for changes.

3 RESULTS

3.1 *Accuracy in image plane*

The influence of the image compression can be expressed by the RMS computed from image coordinates differences relatively to the reference TIFF 16 bit (lossless) variant, which represents the maximum quality. The first image taken in basic center position of the camera with image plane parallel to the reference plane was used for this purpose.

There were 5 targets of the same size for every "target size" variant and 6 targets for every "target shape" variant. If we assume the same accuracy in x and y direction of image coordinates, we can compute the RMS from 10 coordinate differences for 8 "size-tests" (from 7 to 41 pixels) and from 12 coordinate differences for 10 "shape-tests" (from 0.9 to 0.3 b/a ratio, where 1 means a circle).

The comparison of accuracy loss for the maximum image quality achievable in various image formats is visible in the Table 1. The influence of target's shape is visible in the Table 2.

As we can see, the accuracy loss caused by image compression is almost negligible for all tested formats, if we use the maximum quality setting. The targets size influenced results

Table 1. Accuracy loss (pixels) of file formats in their max. quality depending on different sizes of targets.

| File format (QF) | Target size in pixels | | | | | | | Rel. accuracy loss* |
	7	12	22	26	31	36	41	(for 26 pix targets)
TIFF 8-bit	0.005	0.004	0.005	0.002	0.003	0.003	0.005	1:2 650 000
TIFF 8-bit (LZW)	0.005	0.004	0.003	0.002	0.007	0.004	0.008	1:2 650 000
JPEG (100)	0.003	0.003	0.002	0.002	0.008	0.004	0.007	1:2 650 000
JP2 (lossless)	0.002	0.001	0.002	0.001	0.002	0.002	0.010	1:4 100 000
JP2 (100)	0.006	0.003	0.003	0.002	0.009	0.005	0.010	1:5 300 000

*The relative accuracy loss is expressed to the longest dimension of the object projected on image plane (5300 pixels).

Table 2. Accuracy loss (pixels) depending on different shapes of targets.

| File format (QF) | Target shape (b/a ratio) | | | | | | | | | |
	0.9	0.85	0.8	0.75	0.7	0.57	0.5	0.45	0.4	0.3
TIFF 8-bit	0.003	0.003	0.002	0.002	0.003	0.003	0.002	0.003	0.010	0.007
TIFF 8-bit (LZW)	0.002	0.003	0.002	0.002	0.003	0.003	0.004	0.003	0.009	0.004
JPEG (100)	0.004	0.003	0.002	0.003	0.003	0.003	0.003	0.005	0.009	0.008
JP2 (lossless)	0.000	0.001	0.001	0.001	0.002	0.002	0.002	0.002	0.006	0.003
JP2 (100)	0.003	0.003	0.002	0.003	0.004	0.004	0.004	0.005	0.009	0.010

from TIFF variants very slightly. In fact, both of the tested TIFF variants are lossless, only the LZW produces a smaller file. There still was a little difference between TIFF-results caused by the independent conversion from TIFF 16-bit to 8-bit and 8-bit (LZW) and the related loss of information. Best results were obtained for targets with 26 pixels in diameter independent on image format and a recognizable influence of targets shape begins with b/a ratio of 0.4. Overall best results were achieved using the JPEG 2000 lossless variant of image.

The results from the complete test of wide variety of JPEG baseline compression rates is visible in the Figure 3 in a form of a 3D chart.

The achieved RMS values confirm the results from other works—the impact of JPEG compression on the photogrammetric accuracy isn't significant to the compression rates of up to 5%. However for targets smaller than 10 pixels in diameter we can expect the down-grade of accuracy faster than using the targets of 15–30 pixels in diameter.

Interesting results were obtained also using the various compression rates for the JPEG 2000 format (Figure 4).

The accuracy loss is significantly smaller than by the JPEG baseline standard for the compression rates of 30% to 50%. The differences between this two formats are negligible to the compression rates of up to 10%. But if we take into account the fact, that the file size in JP2 format is significantly smaller than the standard JPEG (Figure 5), it is worth considering the more frequent usage of this file format in photogrammetric applications. The JP2 file is still 5-times smaller than the standard JPEG in the maximum quality.

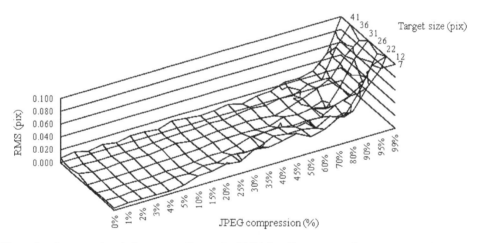

Figure 3. Accuracy loss in image coordinates for JPEG baseline compression rates depending on targets size.

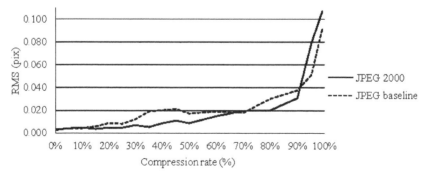

Figure 4. Comparison of JPEG baseline and JPEG 2000 accuracy loss using various compression rates (for targets with 22 pixels in diameter).

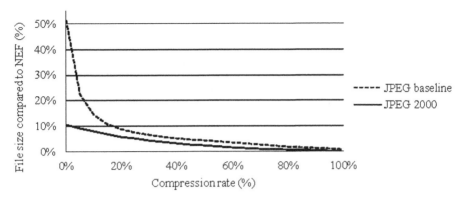

Figure 5. Comparison of file sizes achieved by different JPEG standards compressions relatively to the original NEF raw image file.

Table 3. Accuracy loss in 3D reference coordinate system.

File format (QF)	Compr. (%)	s0	Overall RMS vector length (µm)	RMS CP (µm)	RMS ChckP (µm)	File size/ RAW (%)	Relative accuracy loss*
TIFF 8bit	none	0.262	9.3	0.7	1.9	305.5	1:1 180 000
TIFF 8bit (LZW)	LZW	0.278	9.9	0.6	1.7	186.9	1:1 270 000
JP2 (lossless)	lossless	0.262	9.3	0.3	0.9	76.2	1:2 400 000
JP2 (100)	0%	0.269	9.5	0.9	1.9	10.4	1:1 130 000
JPEG (100)	0%	0.265	9.4	0.4	1.7	51.3	1:1 260 000
JPEG (95)	5%	0.265	9.4	0.7	2.0	22.3	1:1 090 000
JPEG (90)	10%	0.277	9.8	1.1	3.3	14.3	1:660 000
JPEG (70)	30%	0.280	9.9	1.6	4.4	6.5	1:500 000
JPEG (50)	50%	0.292	10.4	2.0	5.3	4.3	1:410 000
JPEG (30)	70%	0.332	11.8	3.5	7.2	2.7	1:300 000
JPEG (10)	90%	0.540	19.2	6.9	10.7	1.2	1:200 000
JPEG (1)	99%	1.650	65.7	18.5	40.7	0.9	1:50 000

*The relative accuracy loss caused by image compression is based on the RMS computed from check points and the objects dimensions, which is approximately 2200 mm in the longest direction.

3.2 *Accuracy in 3D reference coordinate system*

The multi-image convergent photogrammetry processing in PhotoModeler was performed with the same calibration file for all compression variants. The average angle of intersection was 63 degrees, average number of rays per 3D point was 10. The 3D reference coordinate system was defined in the first TIFF 16-bit variant, the 3D coordinates of 8 control points were exported and used to transform every other variant into the same 3D coordinate system as the first.

The quality of 3D photogrammetric processing can be evaluated in different ways. We can observe the values of sigma 0 after bundle adjustment and the points precision estimated by the software (Overall RMS vector length), or we can compare the final 3D coordinates of control points (CP) and check points (ChckP) with the reference coordinates (Table 3).

As we can see in Table 3. Best results were obtained using the Wavelet-based JP2 format with lossless compression. The accuracy loss in standard JPEG baseline format is also negligible to the 5% compression rate and comparable to TIFF 8-bit results. The theoretical accuracy of multi-image convergent photogrammetry can be estimated using the following formula (Fraser, 1984):

$$m_p = \frac{M_s \cdot m' \cdot q}{\sqrt{k}} \qquad (1)$$

where the M_s represents the image scale number (77 in our case), m' is an accuracy of image coordinate measurement (0.49 μm in our case), q is a quality factor describing camera configuration (0.5–1.2) and k is average number of photos taken from one camera position. The expected accuracy of points in ref. coordinate system equals 0.02 mm which corresponds with the obtained results.

4 CONCLUSION

The results show that JPEG baseline and JPEG 2000 lossy compressions can be used in high accuracy 3D photogrammetric processing with quality factors 95–100 with no worry about significant accuracy loss. The JPEG 2000 format yet seems to be more effective in the sense of quality/size ratio. The implemented JPEG support in DSLRs can be sufficient, if we use the maximum quality JPEG variant (JPEG fine setting in Nikon cameras), which compression rate should not reach the 5% rate limit (based on the file sizes). This is interesting e.g. in applications with the need of online processing and direct transfer of images to a computer, where the big file sizes and NEF-format could complicate the process.

However, in special applications such as time baseline method, where we analyze a sequence of images taken from one stable camera position, we deal with the need of subpixel accuracy in rates of 0.01–0.001 pixels. In this cases it seems appropriate to shoot in RAW and convert images by JPEG 2000 lossless compression. The relative accuracy loss in image coordinates could then reach value of 1:4 000 000 (considering the use of big enough targets and LSM method for target center estimation).

ACKNOWLEDGEMENTS

This article was created with the support of the Ministry of Education, Science, Research and Sport of the Slovak Republic within the Research and Development Operational Programme for the project "University Science Park of STU Bratislava", ITMS 26240220084, co-funded by the European Regional Development Fund.

REFERENCES

Akcay, O., Erenoglu, R.C. & Avsar, E.O. 2017. *The effect of JPEG compression in close range photogrammetry*. International Journal of Engineering and Geosciences (IJEG). Vol 2, Issue 1, pp. 35–40, February, 2017, ISSN 2548-0960, Turkey.

Fraser, C.S. 1984. *Network design optimization in non-topographic photogrammetry*, IA PRS, Vol. XXV, part A5, pp. 296–308, Rio de Janeiro, 1984.

Jaakola, J. & Orava, E., 1994. *The Effect of Pixel Size and Compression on Metric Quality of Digital Aerial Images*. International Archives of Photogrammetry and Remote Sensing, Vol. 30, Part 3/1, pp. 409–415.

Kiefner, M. & Hahn, M. 2000. *Image compression versus matching accuracy*. International Archives of Photogrammetry and Remote Sensing. Vol. XXXIII, Part B2. Amsterdam.

Luhmann, T., Robson, S., Kyle, S., et al. 2013. *Close-Range Photogrammetry and 3D Imaging*. Berlin, Boston: De Gruyter. Retrieved 25 Jun. 2017.

Šedina, J. & Pavelka, K. 2016. *Precise Photogrammetric Methods for Deformation Measurements*. Interdisciplinarity in Theory and Practice. 2016(10), s. 285–290. ISSN 2344-2409.

Urban, R., Braun, J. & Štroner, M. 2015. *Precise deformation measurement of prestressed concrete beam during a strain test using the combination of intersection photogrammetry and micro-network measurement*. Proceedings of SPIE The International Society for Optical Engineering, 9528, pp. ISSN 0277-786X.

Advances and Trends in Geodesy, Cartography and Geoinformatics – Molčíková et al. (Eds)
© 2018 Taylor & Francis Group, London, ISBN 978-1-138-58489-1

Photogrammetric deformation measurement of concrete flat slab

M. Marčiš, M. Fraštia & T. Augustín
Faculty of Civil Engineering, Slovak University of Technology, Bratislava, Slovakia

ABSTRACT: Measuring of building components deformations during load tests is mostly done by the methods of length difference measurement, using instruments like dilatometers and inductive sensors. But during this measurements problems can occur related to contact measurement, measurement of only one dimension (length), limited number of sensors located on the object, need of cabling, influence of other physical environment properties (humidity, temperature) etc. These complications can be well eliminated by the non-contact optical methods. In our contribution we use various methods of digital photogrammetry for the documentation of static deformations in various load stages, specifically a combination of time baseline and multi-image convergent photogrammetry. The main goal of this measurements was the determination of deformation in the plane of the concrete surface expressed by length changes between observed points. This was possible thanks to calculating the changes of image coordinates of points using the time baseline method. Accuracy at sub-millimeter level was achieved and advantages of this measurement technique are commented.

1 INTRODUCTION

Digital photogrammetry is widely used in civil engineering most commonly either as a method for documenting the current state of the object (Pukanská et al., 2014; Bíla, Šedina & Pavelka, 2014; Kliment & Halva, 2009) or to track its movements and deformations (Šedina, Pavelka & Housarová, 2016). The multi-image convergent photogrammetry clearly belongs to the currently most accurate methods of digital photogrammetry used to reconstruct the 3D shape of measured objects. Method's main advantages are especially the considerable freedom in camera positions and relative orientation. With the use of circular targets, accurate measurement of image coordinates in tenths and hundredths of a pixel can be achieved which, depending on camera distance to object and camera used can be translated into a high attainable accuracy in the reference coordinate system—in tenths and hundredths of a millimeter (Luhmann, 2006; Urban, Braun & Štroner, 2015). In special cases, simpler methods of digital photogrammetry can be used—such as time base-line method. The use of this method allows us to determine displacements of observed points in a reference plane parallel to the image plane. If the observed changes are small enough, it is possible to neglect the calibration of the camera and make the processing even simpler.

In the first step 3D coordinates of observed points were determined by the multi-image convergent photogrammetry. These coordinates served to control the single camera position and orientation in the zero-load stage relatively to the plane of the concrete panel and to determine the image scale for the purpose of time base-line method.

2 SUBJECT OF MEASUREMENT

Load-test of a composite steel-concrete panel with symmetrically placed openings from the column face to the edge of the flat slab (Figure 1 left) was the subject of photogrammetric measurement presented in this paper. An octagonal segment of flat slab with spans 5400 mm × 5400 mm (size of tested specimen was 2400 mm × 2400 mm) was the specimen.

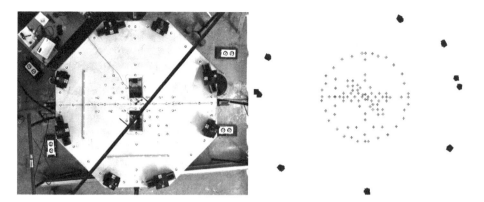

Figure 1.   Observed concrete specimen (left) and camera positions configuration in the first load-stage (right).

Thickness of analyzed flat slab was 250 mm, the dimensions of supporting column 200×200 mm ($4\phi12$). During the experimental works, the specimen was loaded with hydraulic jack situated under the column. The specimen was tied to strong floor of laboratory with 8 steel rods. This contribution describes only one specimen from a series of flat slabs with different reinforcement, while the aim of this experiments in building point of view is to determine the influence of holes in flat slabs.

In this case, the convergent photogrammetry was used for the documentation of the first static stage with no pressure (Figure 1 right). 3D coordinates of observed points were determined with high spatial accuracy, better than 0.1 mm. Then the time base-line method with only one stable camera (Figure 1 left) was used to measure the changes of image coordinates of observed points on the concrete panel in every stage of the load test. The main goal of photogrammetry was to determine the relative distance changes between neighboring points.

3   ACQUIRING DATA DURING LOAD TEST

3.1   *Multi-image convergent photogrammetry*

84 PhotoModeler RAD targets were glued on the surface of the concrete panel. Another 8 RAD-targets were stabilized on the ground around the load test construction. This 8 targets served as points for verification of camera fixed position.

Scale was determined using 2 scale bars attached on the concrete surface—the length of scale bars was derived from calibration plate of the Comet L3D triangulation scanner with an accuracy of 0.1 mm. 16 images have been taken using a DSLR camera Nikon D800E (full-frame 36 Mpixel sensor) with a Nikkor 35 mm AF-S ED 1:1.8G lens.

Images have been taken from a ladder around the panel from 8 stations, 2 images on every station (one in normal position and one rotated 90 degrees around the axis of view). To eliminate the blur motion caused by bad illumination conditions an external flash was used. Exposition time of 1/250 s was achieved at the following camera settings: aperture priority—F/8, ISO 200, autofocus was used because of varying camera-to-object distance.

3.2   *Time base-line method*

The images for the time base-line method were taken also by the Nikon D800E, but with different lens—SIGMA 50 mm 1:1.4 DG, because of the size of the field of view and the position of the camera relative to the concrete panel. The camera was stabilized on a crane located over the load test experiment in 6 m height over the flat slab. Perpendicular orientation and center position of camera with regard to the concrete panel was achieved using a plumb line.

The camera was operated from the ground using a wireless remote control to minimize the impact of shaking during manual shutter release.

After achieving a current load state, the values of stab deformations measured by digital dial gauges were recorded and relative displacements between special metal markers were manually measured using a mechanical deformeter. A set of 3 images was taken to verify the repeatability of image coordinates measurement immediately after that. Whole load test consisted of 10 load stages. The concrete panel mechanically collapsed immediately after achieving the load value of 700 kN.

4 IMAGE PROCESSING

Processing of images by the method of multi-image convergent photogrammetry was done by PhotoModeler software.

The a-priori accuracy of convergent photogrammetry is possible to estimate using the following formula (Fraser, 1984):

$$m_p = \frac{M_s \cdot m' \cdot q}{\sqrt{k}} \tag{1}$$

where the M_s represents the image scale (168.2 in our case), m' is an accuracy of image coordinate measurement (0.49 µm in our case), q is a quality factor describing camera configuration (0.5–1.2) and k is average number of photos taken from one camera position (2). The expected accuracy of points in reference coordinate system equals 0.05 mm which corresponds with the obtained results (Table 1).

High accuracy in photogrammetry processing is achievable only thanks to quality camera calibration (Urban, 2008), therefore on the job calibration was done on the objects points which led to maximum residuals under 1 pixel.

All 3D points from convergent photogrammetry were used as control points (CPs) for the single-image processing in corresponding 0-load stage. Using the "control solution" option in PhotoModeler SW the exterior orientation parameters of the camera were solved. Maximum residual on points after bundle adjustment was smaller than 0.4 pixel. The principle ray of autocollimation of camera deviated from the perpendicular direction to the concrete panel surface only 1 degree (the image plane was almost parallel to the reference plane). The projection center of the camera deviated 60 mm from the center of the concrete panel. Based on the measured distances between points in reference coordinate system and in the image coordinate system it was possible to calculate the size of pixel projected on the panel surface by value of 0.576 mm. At this value and supposed accuracy of 0.1 pix of measured image coordinates, we can assume reference accuracy of measured length:

$$m_l = GSD \cdot 0.1 \cdot \sqrt{2} \approx 0.08 \, mm \tag{2}$$

This accuracy could by theoretically better if we assume that the repeatability test based on 3 images taken in every load test stage produced differences of image coordinates on identical points with RMS = 0.03 pix.

Table 1. Statistics from 0-load stage of multi-image convergent photogrammetry.

Average nr. of rays per point	Average nr. of points per photo	S0	Max. residual pix	RMS pix	Aver. inters. angle °	Aver. object distance m	Overall RMS in ref. coord. sys. mm
8	69	0.885	0.590	0.112	82	3.1	0.03

The processing of time base-line photogrammetry was carried out again in PhotoModeler SW, but only for the automatic measurement of image coordinates which have been exported and next computed its changes. As the results showed, there was no need to eliminate radial (optical) distortion of lenses using a precise camera calibration in this step. Overall 84 images were processed.

5 ANALYSIS AND EVALUATION OF THE RESULTS

As mentioned before, there were 8 targets stabilized on the ground around the tested concrete segment. This twins of points in 4 corners of the images served to analyze the stability of the camera during the load test (Figure 2).

The charts in Figure 2 describe the image coordinate differences of control points from the "30 kN" load stage. The coordinates of all 8 targets changed in the same way with differences not bigger than 0.06 pixels each other, therefore only the average values are shown in the charts. Based on detected changes of image coordinates it is possible, that the camera system slightly tilted during the load test. However, the relative distances between the upper left and right points have changed max. up to 0.03 pixels. So we can assume that the instability of the camera and change of scale were not significant.

In the next step distances between neighboring points on the concrete panel were computed and the changes of distances were analyzed. This changes in following load stages showed a significant influence of 3D surface deformation in vertical direction (flexure of panel) and caused impossible negative differences between distances (as if the outer lengths in radial direction became shorter). The influence of the central projection in combination with panel deformation is visible on Figure 3, where Δh represents vertical displacement of points from the reference plane.

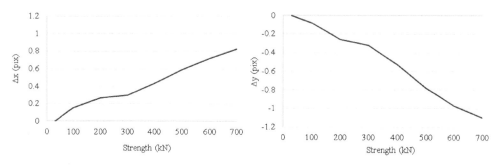

Figure 2. Average movement of ground control points in the image coordinate system (changes in pixels – Δx left, Δy right).

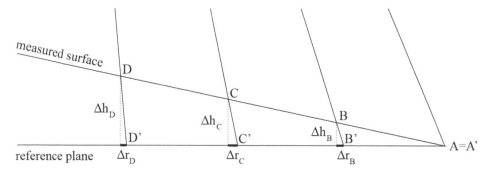

Figure 3. Radial displacements of image coordinates caused by central projection.

86

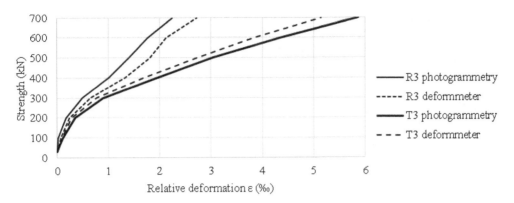

Figure 4. Comparison of radial (R) and tangential (T) relative deformation between points in 200 mm grid in a 100–300 mm distance from the slabs center.

From Figure 3 we assume that the length AB = BC = CD. The projected length A'B' is definitely shorter than the original AB. If we correct these lengths using the radial displacements Δr we obtain orthogonal projections of the lengths AB, BC, CD into the reference plane. The angle between the reference surface and the measured surface is in the maximum deformation state so small, that we can assume the length AB ≈ A'B' + Δr$_B$.

Digital dial gauges installed under the concrete stab measured vertical deformations Δh of the stab. We used these values as the input to calculate the corrections Δr of radial distances between the center of the concrete panel and the selected observed point. Every point in every load stage was corrected this way. Finally, it was possible to calculate relative deformation (ratio of changed and original length in ‰) of every distance between selected points in radial and also in tangential direction. The growth of the relative deformation in selected areas was compared with the values from mechanical deformeter measurement (Figure 4).

The differences between the results from these two methods have a systematical character caused most likely by the insufficient camera stability and its slow tilting during the 1.5-hour test (projective distortion). However, we deal here with differences in maximum load stage of 0.5 ‰ on a 200 mm length, which equals 0.1 mm. Assuming a better camera stability we could obtain more accurate results.

6 CONCLUSION

The digital photogrammetry is known as an effective and accurate method for displacement measurements during load-tests of construction elements and it enables a total control not only of the points on the tested object but also on the load-test construction around it. The results from such measurements create a complex image of the deformations of the observed object and, after a deep analysis, it can help to adjust the whole structure of the experiment removing unreliable elements.

This approach can be made even more effective using time base-line method in the sense of processing simplicity, if we need to evaluate the points displacements only in a reference plane (2D). However, to obtain reliable results we need to calculate also with the possibility of spatial deformation of the observed surface. These 3D deformations can be evaluated using additional equipment such as digital dial gauges or measuring and processing the complete load test using multi-image convergent photogrammetry.

The main advantage of the time base-line method combined with the digital dial gauges is, that there is no need to measure the length between points manually using a mechanical deformeter and it makes the load-test progress faster. Also there is no limit in number of observed points that can be fixed on the concrete surface.

However, to achieve the high precision and reliability of the measurement it is needed to pay special attention to the stabilization of the reference coordinate system, definition of the scale, signalization of the observed points, configuration and stabilization of camera and finally processing parameters too.

ACKNOWLEDGEMENTS

This article was created with the support of the Ministry of Education, Science, Research and Sport of the Slovak Republic within the Research and Development Operational Programme for the project "University Science Park of STU Bratislava", ITMS 26240220084, co-funded by the European Regional Development Fund.

REFERENCES

Bílá, Z., Šedina, J., Pavelka, K., 2014: *Spatial documentation and visualisation of historical artifacts*, International Multidisciplinary Scientific GeoConference Surveying Geology and Mining Ecology Management, SGEM, 2014, Vol. 3, no. 2, pp. 299–306. ISSN 1314-2704.

Fraser, C.S., 1984: *Network design optimization in non-topographic photogrammetry*, IA PRS, Vol. XXV, part A5, pp. 296–308, Rio de Janeiro, 1984.

Kliment, M., Halva, J., 2009: *Photogrammetric data utilization in vertical thematic mapping in land consolidation project*, Reminiscencie geodézie a fotogrametrie, [CD-ROM], odborný seminár s medzinárodnou účasťou, Bratislava 16. jún 2009., Bratislava: STU, ISBN 978-80-227-3045-7.

Luhmann, T., 2006: *Close Range Photogrammetry*. Whittles Publishing, Scotland, UK, 2006. 510 s. ISBN 1-870325-50-8.

Pukanská, K., Bartoš, K., Weiss, G., Rákay, Š. ml., 2014: *The Application of close-range photogrammetry for a documentation of metallurgical art-historical objects*, SGEM 2014, 14th international multidiscilinary scientific geoconference: GeoConference on Informatics, Geoinformatics and Remote Sensing, conference proceedings, 17–26 June, 2014, Albena, Bulgaria. – Sofia, STEF92 Technology Ltd., P. 327–334., ISBN 978-619-7105-12-4.

Urban, R., 2008: *Lens Distortion Influence Suppression*, Geodézia, kartografia a geografické informačné systémy 2008 [CD-ROM], Košice: Technical University BERG Faculty, p. 1–10. ISBN 978-80-553-0079-5.

Urban, R., Braun, J., Štroner, M., 2015: *Precise deformation measurement of prestressed concrete beam during a strain test using the combination of intersection photogrammetry and micro-network measurement*, Proceedings of SPIE The International Society for Optical Engineering, 9528, pp. ISSN 0277-786X., SCOPUS.

Šedina, J., Pavelka, K., Housarová, E., 2016: *Using of photogrammetric methods for deformation measurements and shape analysis*. In: Advances and Trends in Engineering Sciences and Technologies II: Proceedings of the 2nd International Conference on Engineering Sciences and Technologies, 29 June – 1 July 2016, High Tatras Mountains, Tatranské Matliare, Slovak Republic. 2nd International Conference on Engineering Sciences and Technologies. Tatranské Matliare, 29.06.2016 – 01.07.2016. Boca Raton: CRC Press, s. 841–846. ISBN 9781138032248.

Advances and Trends in Geodesy, Cartography and Geoinformatics – Molčíková et al. (Eds)
© *2018 Taylor & Francis Group, London, ISBN 978-1-138-58489-1*

High-resolution 3-D mapping for the survey of valuable inaccessible Medvedia Cave in the National Park of Slovenský raj

K. Pukanská, K. Bartoš & J. Sabová
Institute of Geodesy, Cartography and GIS, Technical University of Košice, Košice, Slovakia

D. Tometzová
Institute of Earth Resources, Technical University of Košice, Košice, Slovakia

ABSTRACT: The objective of this paper is a geodetic measurement of the Medvedia (Bear) Cave, the largest and most beautiful cave in the National Park of Slovenský raj. The cave itself is part of the cave complex under the Glac plateau. This cave filled with beautiful stalagmites, stalactites and other cave formations was created by river erosion in tribasic limestone. The whole complex is situated in the Nature Reserve of Kysel. The entrance to the cave is situated at an altitude of 905 m asl.

The survey was realised by terrestrial laser scanner Leica ScanStation C10. Subsequently, its spatial model, individual cross-sections as well as the overall map of this publicly inaccessible cave were created. The cave is hardly accessible and unlighted. About 300 m long part of the cave with wider corridors and remains of the cave bears was the subject of the survey. The unique high-resolution 3D digital model, used mainly for the presentation of this inaccessible cave in tourism, is the main result of this geodetic measurement.

1   INTRODUCTION

1.1   *Speleological mapping*

Karst and cave systems are complex three-dimensional phenomena. Their mapping and visualisation present a challenge in cartography, morphology, but also computer graphics. Creation of plans of cave objects or spaces, generation of analogue and vector maps, and also the creation of 3D digital cave models is an important part of cave exploration and research (Gallay, 2016). Cave maps and their three-dimensional models provide a more detailed explanation and insight into the tectonic structure of the given area for geologists and hydrologists (Bella, 2016).

Speleological mapping, or surveying, is a set of activities related to the determination of spatial relations in a speleological environment. Regarding geodetic and cartographic methods, mostly methods of mining surveying, engineering surveying and photogrammetry are used in speleological mapping. Mapping of speleological areas addresses the issue of creating speleological maps. For this process, it is necessary to collect the existing usable materials, their completion, preparations for measurements, establishment or completion of geodetic point fields on the surface and their connection with underground spaces, subsequent detailed survey and necessary computational and graphical works for the creation of a map.

1.2   *Site description*

The national natural monument Medvedia (Bear) Cave is hidden on the edge of the Glac plateau in the National Park of Slovenský raj (Fig. 1). It is one of the largest and most beautiful caves in Slovenský raj and one of the most important paleontological sites. It can be characterised by great natural and aesthetic values. There is a rich occurrence of Ursus

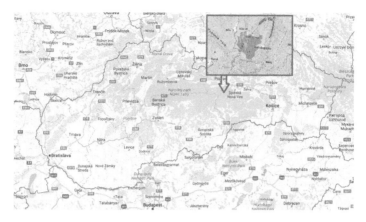

Figure 1.   Location of the Medvedia Cave in the National Park of Slovenský raj.

Figure 2.   Decoration of the Medvedia Cave.

Spelaeus bones, a wide variety of wintering bats, the presence of paleogeneous rocks and phosphorus minerals, but also wide range of cave fillings. The 12-metre wide entrance to the cave, which is situated at the altitude of 905 m asl., gradually narrows into a tight 6 m long corridor (Fig. 3). Through this area, we can get into the Dome of Bats, which is the part of about 2 m wide and 25 m high.

The cave was discovered and explored by members of the Slovak Speleological Society from Spišská Nová Ves in 1952. Approximately 497 m of the cave is mapped with the elevation of about 30 m. The lowest point of the cave is at the entrance, and the highest point (933 m asl.) is in the northern part of the Dome of Dionýz Štúr. The cave was created by an underground stream, as indicated by oval shapes with significant lateral channels and rounded widths in clay sediments. Underground water flowed to the cave from the meadowy plateau Veľká Poľana.

The most remarkable part of the cave is the 20 m wide and 40 m long Palm Hall. It has a unique dripstone decoration of yellowish colour, sinter ponds, palm groves and stalactite decoration (Fig. 2). Stalagmites reaching up to 3 metres height, but also straws of various lengths, are present to the greatest extent.

The colour spectrum of cave fillings is very diverse, from white to yellowish and light brown, being dominant for the youngest formations, while older formations have a red-brown colour. The composition of cave sediments is also very varied. There are sediments of sinter character with fragments of light ocherous and finely grained sandstone with scattered parts of limonite concretes. The cave consists of the following parts: 1. Entrance Corridor, 2.

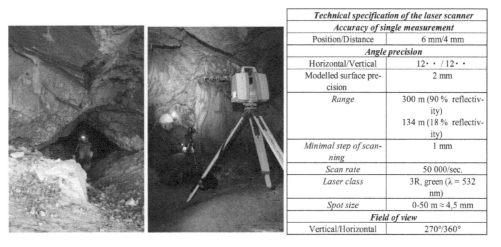

Technical specification of the laser scanner	
Accuracy of single measurement	
Position/Distance	6 mm/4 mm
Angle precision	
Horizontal/Vertical	12ʺ / 12ʺ
Modelled surface precision	2 mm
Range	300 m (90 % reflectivity) 134 m (18 % reflectivity)
Minimal step of scanning	1 mm
Scan rate	50 000/sec.
Laser class	3R, green (λ = 532 nm)
Spot size	0-50 m ≈ 4,5 mm
Field of view	
Vertical/Horizontal	270°/360°

Figure 3.   The entrance to the cave and the image of geodetic activities in the cave and technical specification of the scanner.

Crossroads, 3. Gorges, 4. Dome of Dionýz Štúr, 5. Hall of the Discoverers, 6. Short Hall, 7. Pond Hall, 8. Palm Hall, 9. Labyrinth, 10. Bear Cemetery, 11. Charnel House, 12. Crooked Corridor, 13. Hall of Slovak Speleologists.

2   MAPPING OF THE CAVE BY TERRESTRIAL LASER SCANNING

2.1   *Preparatory works and geodetic measurement of the cave by terrestrial laser scanning*

Preparatory works consisted of studying the provided map data from 2003, which were available from an earlier geodetic measurement (Fig. 4). The map is created in the national coordinate system of the Slovak Republic—Datum of Uniform Trigonometric Cadastral Network and vertical datum—Baltic Vertical Datum—After Adjustment at a scale of 1:750. The mapping was realised for the needs of dissertation thesis of the co-author. The results of the survey were used to design a new geopark in the area of Lower Spiš, which has the potential for recreational purposes. Resulting visualisations can virtually advance the cave spaces towards visitors of the geopark, for example in a multi-functional 3D cinema.

2.2   *Data collection using terrestrial laser scanner Leica ScanStation C10*

Terrestrial laser scanning and digital close-range photogrammetry are currently the most popular technologies for creating three-dimensional digital models of objects (Majid, 2017). In this case, the geodetic measurement was realised by terrestrial laser scanning, since the light conditions did not allow to use a digital camera to create an optical record. After the terrain reconnaissance and appropriate configuration of scanning stations, we proceeded to the scanning process itself. The measurement was realised from survey stations determining a traverse with spatial resolution 3 cm per 10 m. From a geodetic point of view, it was an open traverse in a local coordinate system, where the Y-axis was defined on the line between points 5001 and 5002. The origin of the coordinate system was defined at the point 5001. In total, laser scanning from 14 survey stations was necessary for geodetic measurement of the whole cave (Fig. 5). The geodetic measurement was based on the principle or measuring the necessary elements for the determination of traverse, i.e. orientations to ground control points, which subsequently served to perform the spatial transformation of individual scans into the common coordinate system. The method of dependent centring was used to achieve higher accuracy. Table 1 shows the quality of orientation at individual scanning stations.

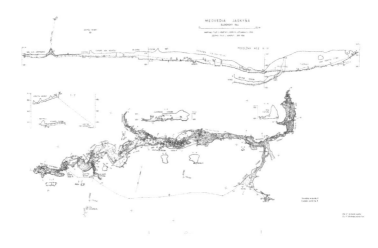

Figure 4.   The original ground plan of the cave and its vertical cross-section.

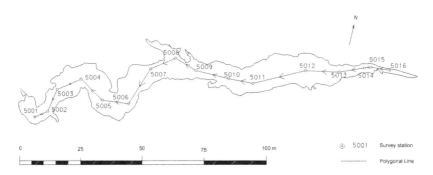

Figure 5.   Scanning stations.

Table 1.   Accuracy of orientation when measuring the traverse using a terrestrial laser scanner.

Station	Orientation	Δx [mm]	Δy [mm]	Δz [mm]	Δd [mm]
5001	5002	0	0	0	0
5002	5001	0	0	−4	0
5003	5002	1	1	−1	1
5004	5003	1	0	3	1
5005	5004	0	0	4	0
5006	5005	0	0	6	0
5007	5006	0	1	4	1
5008	5007	2	0	0	2
5009	5008	0	0	2	0
5010	5009	0	0	5	0
5011	5010	1	1	3	1
5012	5011	2	0	5	2
5013	5012	0	0	4	0
5014	5013	0	0	4	0

2.3   *Data processing and final visualisation in CAD*

The processing of obtained data, which consisted of approx. 53 million of points in the form of the point cloud, was realised by creating the TIN model (Fig. 6). Sometimes, also the point cloud can be considered as the final output of measurement, because it can appropriately

Figure 6. The final TIN model—in the Labyrinth with a view towards the Bear Cemetery.

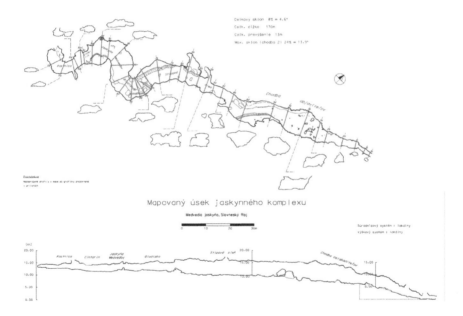

Figure 7. A vertical cross-section through the cave and the ground plan with highlighted positions of cross-sections.

create a spatial image of the given area, which can be additionally complemented by further information in interest segments, for example by dimensioning.

2.4 *Creation of cross-sections and calculation of the volume of the Cave*

The initial objective of the measurement of the Medvedia Cave was not only the measurement and visualisation itself but also the creation of cross-sections of individual parts of the cave and calculation of its volume. First of all, the ground plan was created using CAD software Microstation V8i and Leica Cloudworx (Fig. 7). Large objects, like boulders that are part of the cave, were also highlighted. Finally, technical hachures were used to indicate sudden changes in the course of the ground or minor slopes.

93

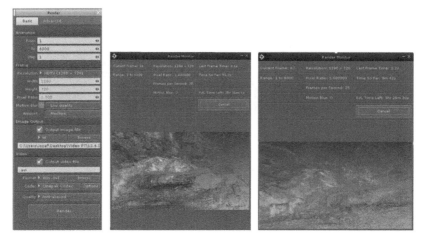

Figure 8.   Creation of video presentation of the original view as well as stereoscopic view in the Pointools software.

2.5   *Video visualisation and presentation*

It was necessary to visualise and present individual results of processing, settings, editing and modelling of data itself. However, the video animation was generated directly from the point cloud, using measured data in the original resolution. The video animation was created using the Bentley Pointools software (Fig. 8). During the creation of video animation, the following properties of the point cloud were adjusted—intensity, colour, compression, import units. Moreover, the software options allow showing the point cloud by stereoscopic techniques. The final result is presented by the video animation, which can also be seen by stereoscopic vision using anaglyphic glasses.

3   CONCLUSION

The main objective of the geodetic measurement and processing of obtained data was not only to get quality and accurate spatial data and create map documentation and cross-sections, but mainly to make the Medvedia Cave more available to the public by a video presentation, since the cave is situated in a hardly accessible terrain, and it is not open to the public. Therefore, its treasures are hidden and can only be seen by a small group of people accompanied by speleologists. The cave is not even illuminated; its visit is only possible with the help of temporary lighting. Modern visualisation technologies and especially terrestrial laser scanning are a really helpful tool for accurate measurement and at least visual accessibility of otherwise inaccessible spaces.

REFERENCES

Bella, P. et al. 2016. Hydrothermal speleogenesis in carbonates and metasomatic silicites induced by subvolcanic intrusions: A case study from the Štiavnické vrchy mountains, Slovakia, *International Journal of Speleology,* Volume 45, Issue 1, January 2016, Pages 11–25.
Gallay, M. 2016. Geomorphometric analysis of cave ceiling channels mapped with 3-D terrestrial scanning, *Hydrology and Earth System Sciences*, Volume 20, Issue 5, 10 May 2016, Pages 1827–1849.
Majid, Z. 2017. Three-dimensional mapping of an ancient cave paintings using close-range photogrammetry and terrestrial laser scanning technologies, *The International Archivers of the Photogrammetry, Remote Sensing and Spatial Information Sciences,* Volume XLII-2/W3, 3D Virtual reconstruction and visualization of Complex Architectures, 1–3 March 2017, Nafplio, Greece.

Advances and Trends in Geodesy, Cartography and Geoinformatics – Molčíková et al. (Eds)
© 2018 Taylor & Francis Group, London, ISBN 978-1-138-58489-1

The influence of refraction on determination of position of objects under water using total station

Š. Rákay, K. Bartoš & K. Pukanská
Ecology, Process Control and Geotechnologies, Faculty of Mining, Technical University of Košice, Košice, Slovakia

ABSTRACT: Refraction is a physical phenomenon that a surveyor encounters relatively often during practical field measurements. Atmospheric refraction, either vertical or horizontal, which is caused by the diffraction of optical properties of the environment by the temperature difference of air layers, is one of the most frequent types of refraction. In non-traditional cases, we can encounter the refraction even when measuring objects under water. In that case, the bending of light when passing through the air/water interface is more substantial. While the distortion caused by atmospheric refraction becomes evident slowly and its effect is noticeable and significant (from a geodetic point of view) up to distances in order of several 10 meters, the effect of distortion in refraction when measuring through water-level is in the order of several mm and detectable up to distances in order of several 10 centimetres. In technical practice, there can sometimes be the case that it is necessary to survey the course of the terrain that continues below the water level. The selection of measurement methodology and used instrumentation depends on the specific conditions and accuracy requirements. When measuring the relief of the bottom by the spatial polar method using total station through the air/water interface, we encounter errors caused not only by refraction but also by the change of the velocity of electromagnetic wavelength propagation of the laser distance meter in a water environment. Therefore, we do not measure the correct position and height of the given point (object), whose image we see above the water level. When determining the horizontal position of points, there are length differences between their real and virtual position. This is in particular related to the change in the velocity of electromagnetic waves passing through the water layer, so the position of measured points is always located further away from the survey station than their actual position. Height differences between the measured and real position can have positive or negative values depending on the size of the measured vertical angle.

1 INTRODUCTION

There are several types of methodologies that can be used to determine the course of shape of a terrain relief below the water level, depending on specific conditions of the given cases, such as the required accuracy of the resulting model, availability of surveying instruments, work efficiency, etc. Typically, acoustic measurements (various types of sonar), airborne laser scanning or airborne laser bathymetry are used for these purposes. Mapping shallow-water depths can be done by total station (TS) with or without surveying prism, or by terrestrial laser scanning (TLS). Coordinates of points determined by TS or TLS have to be corrected for the systematic errors in depth and distance that are caused by the refraction of the laser beam at the interface between air and water. The issue of incorrect determination of points position under the water surface measured by the spatial polar method using terrestrial laser scanning (TLS) is solved in works (Smith et al., 2012; Deruyter et al., 2015).

2 THEORETICAL BACKGROUND

The position of a point in the form $P(\omega,z,D)$ is defined by the spatial polar method, in which the position is determined by measuring polar coordinates, i.e. horizontal angle (ω), zenithal distance (z) or elevation angle (β) and spatial distance (D). From a practical point of view, the position of points is calculated by simple trigonometric relations and expressed as $P(x,y,h)$ in a certain Cartesian coordinate system. Each measurement of variables (angles, distances) in the determination of the Cartesian coordinates of the point $P(x,y,h)$ is realised in a physical environment that affects the results of measurement by its variability. Refraction is one of these aspects.

Measuring by a total station uses, like the TLS, a spatial polar method to determine the position of points in space. An electronic distance meter, with which these instruments are equipped, operates on the principle of transmitting and subsequently receiving the light impulse by a laser diode, i.e. electromagnetic waves of a certain wavelength (Rüeger, 2012). When determining the position of points below the water level, two fundamental sources of errors that affect the measurement result need to be taken into account. One is the refraction of electromagnetic waves at the interface of different air/water density, which causes a change in the measured elevation angle. The second one is the change in velocity (slow down) of propagation of electromagnetic waves passing through water environment, which causes a change in the measured spatial distance.

The effect of refraction of a light beam when passing through environments of varying density is described by Snell's law of refraction:

$$\frac{\sin i}{\sin r} = \frac{n_2}{n_1} = \frac{v_1}{v_2} \tag{1}$$

where $i.$ is the angle of incidence, r is the angle of refraction, n_1 and n_2 are refractive indexes of individual environments, and v_1 and v_2 are velocities of propagation of electromagnetic waves in given environments (Halliday & Resnick, 2011). The position of the point P in relation to the survey station S, without the water layer, is determined by measuring the corresponding variables – spatial distance D and elevation angle β, by which the horizontal distance d and height h are calculated according to equations $d = D\sin\beta$ and $h = D\cos\beta'$ (Fig. 1). When measuring the position of the same point (P) through the water layer, the elevation angle β_1 and spatial distance $D' = D_1 + n_2D_2$ are measured when targeting its imaginary image caused by refraction. Part of the measured distance, which electromagnetic wave passes when passing through the water environment (marked as D_2 in Fig. 2) is multiplied by the index n_2, which takes into account the velocity of propagation of electromagnetic waves in a water environment. The imaginary position of the point P, i.e. P', in relation to the survey station S, is obtained by the calculation of horizontal distance $d' = D'\sin\beta_1$ and height $h' = D'\cos\beta_1$.

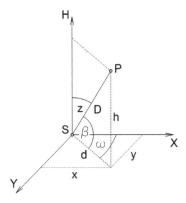

Figure 1. Determination of the position of the point P using cartesian $P(x,y,h)$ and polar $P(\omega,z,D)$ coordinates.

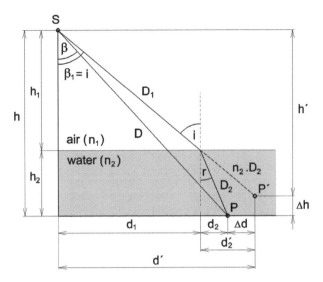

Figure 2. Geometric representation of parameters necessary for the calculation of vertical Δh and length Δd difference in determining the position of points under the water.

Table 1. The calculation of the difference in elevation Δh between the real height of the point P and the height of the point P' depending on the depth of water h_2 and the angle of incidence of electromagnetic wave $\beta_1 = i$ on the water surface.

	Angle of incidence $\beta_1 = i$ [gon]										
	0	10	20	30	40	50	60	70	80	90	
0.1	0.033	0.032	0.030	0.026	0.020	0.011	−0.001	−0.019	−0.041	−0.069	
0.2	0.066	0.065	0.060	0.052	0.040	0.022	−0.003	−0.037	−0.082	−0.138	
0.3	0.099	0.097	0.091	0.079	0.060	0.033	−0.004	−0.056	−0.124	−0.207	
0.4	0.133	0.130	0.121	0.105	0.080	0.044	−0.006	−0.075	−0.165	−0.276	
0.5	0.166	0.162	0.151	0.131	0.100	0.056	−0.007	−0.093	−0.206	−0.345	
0.6	0.199	0.195	0.181	0.157	0.120	0.067	−0.009	−0.112	−0.247	−0.414	
0.7	0.232	0.227	0.211	0.183	0.140	0.078	−0.010	−0.130	−0.288	−0.483	
0.8	0.265	0.259	0.241	0.210	0.160	0.089	−0.012	−0.149	−0.330	−0.551	
0.9	0.298	0.292	0.272	0.236	0.181	0.100	−0.013	−0.168	−0.371	−0.620	
1.0	0.331	0.324	0.302	0.262	0.201	0.111	−0.014	−0.186	−0.412	−0.689	
					Difference in elevation Δh [m]						

Depth of water h_2 [m] (left axis label), Diff. in elevation Δh [m] (right axis label)

*(+Δh: the point P' is vertically below the point P, −Δh: the point P' is higher than the point P).

The difference in elevation Δh between the real position of the point P and the position of the point P' can be calculated as:

$$\Delta h = D \cos \beta - h'$$ (2)

where $h' = (D_1 + n_2 D_2) \cos\beta_1$, $D_1 = h_1 / \cos i$, and $D_2 = h_2 / \cos r$.

The difference in length Δd between the real position of the point P and the position of the point P' can be calculated as:

$$\Delta d = d'_2 - d_2$$ (3)

where $d'_2 = n_2 D_2 \sin i$ and $d_2 = h_2 \tan r$.

Table 2. The calculation of the difference in length Δd between the real position of the point P and position of the point P' depending on the depth of water h_2 and the angle of incidence of electromagnetic wave $\beta_1 = i$ on the water surface.

		Angle of incidence $\beta_1 = i$ [gon]										
		0	10	20	30	40	50	60	70	80	90	
Depth of water h_2 [m]	0.1	0.000	0.009	0.018	0.028	0.038	0.048	0.059	0.070	0.079	0.085	Diff. in length Δd [m]
	0.2	0.000	0.018	0.037	0.056	0.076	0.097	0.118	0.139	0.158	0.171	
	0.3	0.000	0.027	0.055	0.084	0.114	0.145	0.177	0.209	0.237	0.256	
	0.4	0.000	0.037	0.074	0.112	0.152	0.194	0.236	0.278	0.315	0.342	
	0.5	0.000	0.046	0.092	0.140	0.190	0.242	0.296	0.348	0.394	0.427	
	0.6	0.000	0.055	0.111	0.168	0.228	0.291	0.355	0.418	0.473	0.513	
	0.7	0.000	0.064	0.129	0.196	0.266	0.339	0.414	0.487	0.552	0.598	
	0.8	0.000	0.073	0.147	0.224	0.304	0.387	0.473	0.557	0.631	0.684	
	0.9	0.000	0.082	0.166	0.252	0.342	0.436	0.532	0.626	0.710	0.769	
	1.0	0.000	0.091	0.184	0.280	0.380	0.484	0.591	0.696	0.789	0.855	
		Difference in length Δd [m]										

*(+Δd: the point P' is further from the survey station S than the point P, −Δd: the point P' is closer than the point P).

In the following tables, the values of height and length differences depending on the depth of water and the angle of incidence of electromagnetic waves are theoretically calculated according to (2) and (3).

The refractive index n_2 is an important factor characterising the physical properties of the given water environment. Its value depends mainly on the water temperature, the wavelength of the light waves that pass through the water environment and the saturation of the water solution (e.g. salt water/fresh water, etc.). The values of Δh and Δd are calculated for refractive index $n_2 = 1.33147$, which was determined by interpolation according to (Daimon & Masumura, 2007); for distilled water, temperature 20°C and wavelength of $\lambda = 658$ nm of Leica Viva TS15 instrument's electronic distance meter (Leica Geosystems 2011). The value of n_2 can be determined in the field by the measured polar coordinates of the point P (using a surveying prism), polar coordinates of the point P' and the height h_1 or h_2. After the substitution into relations (2) and (3), the value of r is calculated. Using the r, by subsequent substitution into the equation (1), the value of n_2 can be calculated.

3   EXPERIMENTAL MEASUREMENTS

Values of height and length differences of the position of the point P were verified by measurements from three survey stations (Fig. 3). Coordinates of survey stations were determined by total station Leica Viva TS15 with forced centring and coordinates of the point P were determined by using a surveying prism. The point P (Fig. 4) was marked by a retro-reflective survey mark on the bottom of a fountain at a depth of 0.162 m below the water level. A local Cartesian coordinate system with the origin at the point 1 and with the X-axis oriented in the direction to the point 2 was used for spatial determination of coordinates of survey stations and the observed point.

The observed point was located on the line between survey stations No. 1 and 2, the survey station No. 3 was placed approximately perpendicular to this line. Horizontal distances were chosen at approximately the same size, i.e. $d_{1P} \approx d_{2P} \approx d_{3P}$. The survey was realised in two series of measurements with different angles of incidence $i = \beta_1$.

The measurement procedure was as follows: determination of coordinates of survey stations No. 1, 2 and 3; determination of coordinates of the observed point P (using a surveying prism); measurements of the position of the observed point through the water layer at each survey station with the elevation angle $\beta_1 =$ approx. 50gon (in the 2nd series of measurements $\beta_1 =$ approx. 70gon).

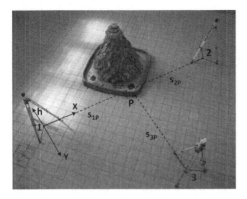

Figure 3. The spatial configuration of survey stations No. 1, 2 and 3 in the measurement of the observed point.

Figure 4. Marking of the point P. In the determination of coordinates using the surveying prism (left), when measuring through the water layer (right).

Table 3. Comparison of results of the 1st series of measurement (for the angle of incidence $\beta_1 =$ approx. 50^{gon}) with numerical values calculated theoretically.

	x [m]	y [m]	h [m]	Angle of incidence $\beta_1 = i$ [^{gon}]	Values calculated according to (2) and (3)	Values calc. from coordinates (x,y,h)
1	100.000	200.000	11.746			
2	103.669	200.000	11.751	P_1' 52.7120	$\Delta h_{1P} = 0.013$ m	$\Delta h_{1P} = 0.014$ m
3	101.829	201.829	11.762		$\Delta d_{1P} = 0.083$ m	$\Delta d_{1P} = 0.084$ m
P	101.833	200.000	10.000	P_2' 52.6012	$\Delta h_{2P} = 0.013$ m	$\Delta h_{2P} = 0.015$ m
P_1'	101.917	200.000	9.986		$\Delta d_{2P} = 0.082$ m	$\Delta d_{2P} = 0.081$ m
P_2'	101.752	200.000	9.985	P_3' 52.3198	$\Delta h_{3P} = 0.014$ m	$\Delta h_{3P} = 0.015$ m
P_3'	101.832	199.917	9.985		$\Delta d_{3P} = 0.082$ m	$\Delta d_{3P} = 0.083$ m

Table 4. Comparison of results of the 2nd series of measurement (for the angle of incidence $\beta_1 =$ approx. 70^{gon}) with numerical values calculated theoretically.

	x [m]	y [m]	h [m]	Angle of incidence $\beta_1 = i$ [^{gon}]	Values calculated according to (2) and (3)	Values calc. from coordinates (x,y,h)
1	100.000	200.000	11.748			
2	106.457	200.000	11.749	P_1' 69.9720	$\Delta h_{1P} = -0.030$ m	$\Delta h_{1P} = -0.030$ m
3	103.227	203.209	11.754		$\Delta d_{1P} = 0.113$ m	$\Delta d_{1P} = 0.113$ m
P	103.243	200.000	10.000	P_2' 69.5215	$\Delta h_{2P} = -0.029$ m	$\Delta h_{2P} = -0.027$ m
P_1'	103.356	200.000	10.113		$\Delta d_{2P} = 0.112$ m	$\Delta d_{2P} = 0.111$ m
P_2'	103.132	200.000	10.111	P_3' 69.3553	$\Delta h_{3P} = -0.028$ m	$\Delta h_{3P} = -0.028$ m
P_3'	103.244	199.888	10.112		$\Delta d_{3P} = 0.112$ m	$\Delta d_{3P} = 0.112$ m

4 EVALUATION OF RESULTS

The results of measurements are summarised in the following tables (Tables 3 and 4). Differences in values calculated according to the theoretical background and equations (2) and (3) are related to the fact that within individual series of measurements, not the same angle of incidence $\beta_1 = i$ was observed at each survey station. Differences between values calculated according to the theoretical background and values calculated from the coordinates (x,y,h) are in the order of 1–2 mm, which could be caused by the accumulation of errors like not knowing the exact value of the refractive index n_2, errors in accuracy of the electronic distance meter, errors in targeting, etc.

Differences in lengths Δd, although they are functions of h_2 and $\beta_1 = i$, are always positive. This is mainly related to the change in the velocity of electromagnetic waves passing through the water layer. Therefore, when measuring distances, the measured distances d are longer than they actually are.

The sign of height differences Δh depends on the angle of incidence $\beta_1 = i$. For values $i <$ approx. 60^{gon} (for refractive index $n_2 = 1.33147$), Δh are positive, i.e. the position of the imaginary point P' is measured lower than the actual height of the point P. For values $i >$ approx. 60^{gon}, the values of Δh has a negative sign, which means that the measured position of the point P' is higher than the actual height h of the point P.

Based on the above theoretical calculation and practical experiment, it can be stated that the accuracy of determining the position of point $P(x,y,h)$ in a certain Cartesian coordinate system depends on the size of length and height differences, which are functions of the water depth and the angle of incidence: Δd, $\Delta h = f(h_2, \beta_1 = i)$. Accordingly, the accuracy of determining the coordinates of the point $P(x,y,h)$ can be defined as:

$$\sigma_x^R = \Delta d \cos \omega, \ \sigma_y^R = \Delta d \sin \omega, \ \sigma_h^R = \Delta h. \qquad (4)$$

where σ_x^R, σ_y^R and σ_h^R are errors of determining the coordinates of the point P measured through the water level using a total station, caused by refraction.

5   CONCLUSION

Surveying the bottom (position of points) under the water by using a total station is one of the possibilities of obtaining spatial information about the course of a given relief. Like any methodology, it has its advantages and limitations. One of the limiting factors is, especially, waviness of the water level caused by the wind, or by water flow, since unpredictable refractions of objects on the bottom can occur on the water level. Another factor is the transparency of water, i.e. an absorption or early reflection of distance meter signal can occur if the water contains soil sediments or algae. The numerical values given in Tables 1 and 2 indicates that relatively large distortions occur during measurement when the accuracy of determination of point position is on the edge of acceptability. According to Table 1, the smallest differences Δh in the determination of heights can be identified for the measured elevation angle $\beta_1 =$ approx. 60^{gon} and the smallest length differences Δd (Table 2) can be identified for the measured elevation angle $\beta_1 = 0^{gon}$. In general, however, height and length differences increase proportionally to the depth of water. Or based on the principles expressed by equations (2) and (3), make a correction of measured variables determining the actual position of measured points by recalculation and thus reconstruct the course of the underwater terrain relief.

REFERENCES

Daimon M. & Masumura A. 2007. Measurement of the refractive index of distilled water from the near-infrared region to the ultraviolet region. *AppliedOptics, 46(18)*: 3811–3820.

Deruyter, G. & Vanhaelst, M. & Stal, C. & Glas, H. & Wulf, A. 2015. The use of terrestrial laser scanning for measurements in shallow-water: Corection of the 3D coordinates of the point cloud. *International Multidisciplinary Scientific Geo Conference – SGEM*:1203–1209.

Halliday, D. & Resnick, R. 2011. Fundamentals of Physics. *9th ed., John Wiley & Sons, Inc., Jearl Walker Cleveland State University, United States of America.*

Leica Geosystems AG 2011. *LeicaTS11/TS15 User Manual, Heerbrugg, Switzerland.*

Rüeger, J.M. 2012. Electronic Distance Measurement: An Introduction. *SpringerScience & BusinessMedia.*

Smith, M. & Vericat, D. & Gibbins, Ch. 2012. Through-water terrestrial laser scanning of gravel beds at the patch scale. *Earth Surf. Process. Landforms 37*: 411–421.

Advances and Trends in Geodesy, Cartography and Geoinformatics – Molčíková et al. (Eds)
© *2018 Taylor & Francis Group, London, ISBN 978-1-138-58489-1*

Enhanced maximal precision increment method for network measurement optimization

M. Štroner, O. Michal & R. Urban
Department of Special Geodesy, Faculty of Civil Engineering, CTU in Prague, Czech Republic

ABSTRACT: Optimization of the measurements is an important theme of the geodesy and especially engineering surveying nowadays. The optimization can be categorized into orders, from the zero to the third order. The zero order design is the coordinate system selection, the first order is the configuration optimization, the second order is the optimization of the number of the repetitions, the third order is the improvement of the existing network by adding point and/or observations. In the engineering surveying the second order design need to be addressed. Thus there are known possible locations of network points, the desired accuracy, possible measurements within the existing situation, and it is necessary to choose such measurements and their numbers of repetition, so that their number was minimal (resp. used energy on their implementation) and yet the required accuracy was met. There were many attempts to meet this objective, but usually by the deployment of general mathematical optimization methods that do not respect the character of geodetic measurements and rational optimization results demands. There was designed a new method which, in our opinion, meets these requirements and is based on gradual selection of the most beneficial measurement to meet the accuracy requirements. The method was successfully tested on the levelling network optimization firstly. Adding of the measurements by one do not allow to use this method successfully on the terrestrial length and angle measurement, because one direction on new station cannot bring precision improvement and thus no new stations are used—except the necessary ones. In this paper was presented newly designed procedure of the optimization using the original principle of the method enhanced by the initial determination based on the least squares method optimization of the whole network, which identifies the initial skeleton for further improvements by the maximal precision increment method. The results of the successful testing of the method on model networks is presented too, including some levelling network (in comparison with the results of the previous method version and brute force perfect solution) and some two dimensional terrestrial networks measured by the total station.

1 INTRODUCTION

Publications (Baarda 1968) and (Grafarend 1974), or later (Grafarend and Sanso 1985) were discussing the topic of the optimization, and laid the fundamentals of optimization in geodesy. An interesting overview of the fundamental principles presents (Amiri-Simkooei et al. 2012).

Algorithms designed to optimize geodetic measurements are aggregated only by adaptations of generally applicable mathematical algorithms that are virtually inapplicable due to the fragmentation of individual measurements and the results in real numbers, although they are mathematically and algorithmically correct, such in (Baselga 2011). Other these algorithms are listes e.g. in (Štroner, 2017a). Already (Baarda 1968) proposed a simple grouping of measurements into logical groups and simple consequent joint increase of the number of repetitions, and since then no other algorithm has been rationally usable. At the article (Štroner, 2017a), a new method was described and the results of testing on the leveling network were shown. At paper (Štroner, 2017b), successful geodetic network optimization tests were shown with independently measured GNSS bases. In the case of terrestrial precise geodetic networks with measured horizontal directions and lengths, however, the application of the basic algorithm fails in some cases because it is not sufficient to add a measurement on a new standpoint in a one-step improvement algorithm, because one measured direction does not contribute to increase the accuracy. This problem occurs especially when the directional accuracy on a given distance (transverse error) is better than the accuracy of the length. Improvement to the method have been proposed to suppress this disadvantage. The basic idea of this algorithm modification is the design of the measurement frame, which is further improved by the basic algorithm.

2 CALCULATION OF THE ACCURACY OF THE ADJUSTED GEODETIC NETWORK

The basic element determining the accuracy of the alignment result is the covariance matrix M describing the accuracy of the resulting coordinates obtained according to (Hampacher, 2015), the adjustment is performed for simplification by the least squares method (in the case of a free network solved by placement conditions the formulas are slightly more complex, but the principle is similar):

$$M = \sigma_0^2 \left(A^T P A \right)^{-1} \tag{1}$$

Here A is the matrix of the plan (Jacobi matrix, derivation matrix) describing the configuration of the geodetic network, P is the matrix of the weights of the individual measurements, in the case of independent measurements, the main diagonal is non-zero only:

$$P = diag\left(p_1 \quad p_2 \quad \cdots \quad p_i \quad \cdots \quad p_m \right). \tag{2}$$

The weight p_i is determined by the chosen constant σ_0 (for simplification it is often chosen equal to 1) and the standard deviations σ_i of the once made i-th measurement l_i and the number of repetitions n_i of the given measurement l_i:

$$p_i = n_i \frac{\sigma_0^2}{\sigma_i^2}. \tag{3}$$

For the purposes of further calculation, it is advantageous (as will become apparent) to separate repeat counts from other variables, the P matrix can be written in the form:

$$P = n \cdot P_0 = diag\left(n_1 \quad \cdots \quad n_m \right) \cdot diag\left(\frac{\sigma_0^2}{\sigma_1^2} \quad \cdots \quad \frac{\sigma_0^2}{\sigma_m^2} \right). \tag{4}$$

Matrix A for m measurements l and k unknowns x (usually determined coordinates):

$$A = \begin{pmatrix} \dfrac{\partial l_1}{\partial x_1} & \dfrac{\partial l_1}{\partial x_2} & \cdots & \dfrac{\partial l_1}{\partial x_k} \\ \dfrac{\partial l_2}{\partial x_1} & \dfrac{\partial l_2}{\partial x_2} & \cdots & \dfrac{\partial l_2}{\partial x_k} \\ \vdots & \vdots & \vdots & \vdots \\ \dfrac{\partial l_{m-1}}{\partial x_1} & \dfrac{\partial l_{m-1}}{\partial x_2} & \cdots & \dfrac{\partial l_{m-1}}{\partial x_k} \\ \dfrac{\partial l_m}{\partial x_1} & \dfrac{\partial l_m}{\partial x_2} & \cdots & \dfrac{\partial l_m}{\partial x_k} \end{pmatrix} \tag{5}$$

Covariance matrix M, which contains the variations on the main diagonal and the corresponding covariances outside the diagonal, and thus provides complete information on the accuracy of the result, is obtained:

$$M = \sigma_0^2 \left(A^T n \, P_0 A \right)^{-1}. \tag{6}$$

3 OPTIMIZATION OBJECTIVES AND CRITERIA

In this case, the ideal optimization goal is to meet the selected accuracy criterion so that the amount of measurement is minimal. The accuracy criterion may be, for example, the maximum standard deviation of the coordinate, the main half axis of the error ellipsoid, the coordinate or positional standard deviations, etc. The criterion is a function of the covariance matrix M. For optimization purposes, all elements of the right side of the formula (6) are constant except the vector n, the accuracy of individual measurements, between which points and point configurations are given by the matrix A and the matrix P_0. Criterion K to assess the fulfillment of the required accuracy is a function of the co-variation matrix:

$$K = f(M) \tag{7}$$

When optimizing, it is necessary to define which K_T criterion is to be met, i.e., the goal is to achieve inequality

$$K \leq K_T \tag{8}$$

4 BASIC PRINCIPLE OF OPTIMIZATION METHOD AND PROPOSED IMPROVEMENT

The principle of the proposed optimization method is that, as the initial state of the calculation, the vector n (repetition of the individual measurements) determines as (virtually) zero, i.e.:

$$n = \left(n_1 \quad \cdots \quad n_m \right) = \left(\varepsilon \quad \cdots \quad \varepsilon \right). \tag{9}$$

Here ε is a very small number close to zero, but so large that it is possible to calculate the inversion in the calculation of the matrix M, while the non-zero number has negligible influence on the calculation of the covariance matrix. It proved to be $\varepsilon = 10^{-12}$. In addition, individual measurements are added to the system in order to improve the worst characteristic as much as possible. This is done as long as the accuracy requirement for all the monitored parameters is not met. Here, for the one-dimensional geodetic networks tested, the method also showed very good results as presented in the publication (Štroner, 2017a) and also in the two-dimensional geodetic networks determined by GNSS (Štroner, 2017b) even when comparing with the global minimum obtained by the calculation of the brute force (by testing all combinations belonging to consideration). When applied to a geodetic 2D network measured with measured horizontal directions and lengths, however, the basic algorithm does not work well in some cases, especially when the accuracy of the directions (on the measurement distance) is metrically better than the length. Then there is practically no chance to set up a new standpoint, and everything is determined the minimum number of standpoints, which is disadvantageous in terms of the number of measurements. To solve this problem, the modification of the calculation procedure was proposed here, where the most important measurements in the network (the skeleton) is firstly determined on the basis of a separate calculation and this frame is then used at the beginning of the calculation.

4.1 Determination of measurement forming a frame

The purpose of frame creation is to set up all standpoints and initiate the optimization direction toward more useful measurements, which ultimately leads to more geodetic solution, i.e. the solution contains more independent redundant measurements and less repetitions of the same measurement. Because it is only about determining the initial state for further optimization, the procedure was designed to be simple. It solves the system of equations here:

$$B \times dn = q - f(n) = b. \tag{10}$$

Here vector x indicates the number of iterations of the intent measurement, i.e., the horizontal direction and the horizontal length, but it is not a vector of integer, but real numbers. The increment of this number is denoted by dn vector, q is the vector of required precision (not covariance matrix, a vector only). The function $f(n)$ then represents a vector describing the achieved accuracy with the current number of measurements in the vector n. Matrix B here includes changes in the accuracy of required accuracy characteristics when changing the number of repeats by one, in the form of variance increment. The difference $q - f(n) = b$ forms the right side of the system of linear equations $B\ dn = b$, whereby before each iteration, the matrix B and vector b must be recalculated. This matrix is obtained by the repeated calculation with the gradual change of the individual number of repetitions. This system is singular, the number of calculated numbers of repetition is usually considerably larger than the number of equations, i.e. the number of coordinates in the geodetic network, the solution can be most easily solved by using Moore-Penrose pseudoinverse (see (Hampacher, 2015)). This is not a correct optimization calculation because it does not minimize the sum of the number of repetitions but fulfill the requirement for accuracy and not from the point of view of compliance (the accuracy achieved is equal or better) but in terms of the difference regardless of the sign.

Additionally, the numbers of repetition are real numbers. The calculation serves only to determine significant measurements and thus the so-called initial frame for further optimization. The boundary of the number of iterations in the vector x, when the measurement is considered to be significant enough for further optimization with initial value 1, is still blurred, it will probably depend on the number of necessary measurements and the achieved repetition counts. For the first algorithm testing, simple rounding was used, i.e. the initial iteration value was 1, if x_i have a value of 0.5 or greater. However, this simplified algorithm development process will definitely be further tested and modified.

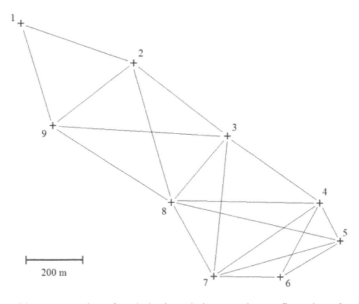

Figure 1. Graphic representation of optimized geodetic network—configuration of points and possible measurements.

5 EXAMPLE OF THE MODIFIED ALGORITHM APPLICATION

The proposed enhanced optimization algorithm has been tested on several geodetic networks with relatively good results. An example will be presented here, Fig. 1 shows the configuration of points and possible measured horizontal directions and lengths, which are considered as one measurement for optimization purposes, and have the same number of repeats.

It is an authentic triangular network of the total dimensions of about $1100 \text{ m} \times 700 \text{ m}$ with 9 points, fixed at points 1 and 5. In total, there is 38 pairs of observations. The required accuracy was given by the maximum major half-axis of error ellipse of 0.75 mm. The accuracy of the directions in the measurement is about 5 times better than the accuracy of the lengths ($\sigma_d = 2 \text{ mm} + 1 \text{ ppm}$; $\sigma_\varphi = 0.1 \text{ mgon}$).

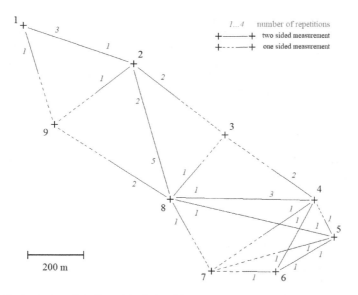

Figure 2.   Optimization result—the original algorithm.

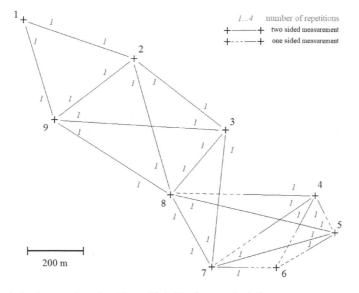

Figure 3.   Optimization result—algorithm with initial frame calculation.

The original algorithm eliminates 3 standpoints and a 15 pairs of the observations, but some observations repeat up to 5x. The total number of measurements is 35. Algorithm supplemented by pre-selection frame will select the observation on all positions except one. The resulting network is much better interlinked, with a lower number of measurements, but with a higher number of used standpoints. The total number of measurements is 30. The results are at Figures 2 and 3.

6 CONCLUSION

The original algorithm based on the maximum precision increment is very simple, easy to apply and provides integer results. It can be applied without problems to one-dimensional problems (egg elevation with measured elevations) or to problems where one measurement (one set, e.g. vector) is not limited by the key dependence on further measurements. If this is not true, the algorithm does not produce the results that are needed and it is necessary to modify it. One way is to create a framework of important measurements in advance. The algorithm shows promising results, as shown in the example above. In addition, it will be necessary to determine more precisely the number of measurements that will form the frame after the preliminary calculation, the testing will continue.

ACKNOWLEDGEMENTS

SGS17/067/OHK1/1T/11 Optimization of acquisition and processing of 3D data for purpose of engineering surveying, geodesy in underground spaces and laser scanning.

REFERENCES

Amiri-Simkooei, A.R.: A new method for second order design of geodetic networks: Aiming at high reliability. Survey Review, 37(293), 2004. ISSN 0039-6265, DOI: 10.1179/sre.2004.37.293.552.

Baarda, W. 1968. A testing procedure for use in geodetic networks. Netherland Geodetic Commission, Delft, Netherlands, 1968.

Baselga, S. 2011. Second order design of geodetic networks by the simulated annealing method. Journal of Surveying Engineering, 137(4), 2011, DOI: 10.1061/(ASCE)SU.1943-5428.0000053.

Grafarend, E.W.: Optimization of geodetic networks. Bolletino di Geodesia a Science Affini, 33(4), 1974.

Grafarend, E.W. - Sanso, F.: Optimization and design of geodetic networks, Springer, Berlin, 1985.

Hampacher, M. & Štroner, M. 2015. Zpracování a analýza měření v inženýrské geodézii. (Processing and analysis of measurement in engineering surveying). 2nd ed. Praha: Česká technika—nakladatelství ČVUT, ČVUT v Praze, 2015. 336 s. ISBN 978-80-01-05843-5. (in Czech).

Štroner, M., Michal, O. & Urban, R. 2017a. Maximal precision increment method utilization for underground geodetic height network optimization. Acta Montanistica Slovaca. 2017, 22(1), 32–42. ISSN 1335–1788.

Štroner, M., Michal, O. & Urban, R. 2017b. GNSS Network Optimization by the Method of the Maximal Precision Increment. In: 17th International Multidisciplinary Scientific Geoconference SGEM 2017 Book 2 Informatics, Geoinformatics, and Remote Sensing Volume II. 17th International Multidisciplinary Scientific Geoconference SGEM 2017. Albena, 2017.

Advances and Trends in Geodesy, Cartography and Geoinformatics – Molčíková et al. (Eds)
© *2018 Taylor & Francis Group, London, ISBN 978-1-138-58489-1*

Errors of electronic high precision short distance measurement

M. Štroner, J. Braun, F. Dvořáček & R. Urban
Department of Special Geodesy, Faculty of Civil Engineering, CTU in Prague, Czech Republic

ABSTRACT: In modern industrial geodesy, high demands are placed on the final accuracy, with expectations currently falling below 1 mm. The measurement methodology and surveying instruments used have to be adjusted to meet these stringent requirements, especially the total. That is why is very important determination of the real accuracy of measurements at very short distances (5–50 m) because it is generally known that this accuracy cannot be increased by simply repeating the measurement because a considerable part of the error is systematic. This article describes the detailed testing of electronic distance meters to determine the absolute size of their systematic errors, their stability over time, their repeatability and the real accuracy of their distance measurement. Based on the experiments' results, calibration procedures were designed and a special correction function for each instrument, whose usage reduces the standard deviation of the measurement of distance by at least 50%.

1   INTRODUCTION

Electronic rangefinders (EDM) have an irreplaceable role in modern terrestrial geodesy. Most geodetic tasks are based on direct distance measurement, and the accuracy of the results is largely dependent on the correct operation of the rangefinder. Since its inception in the 1940s, distance meters have developed a great deal, and large-scale devices have become miniature components of total stations, which, unlike a man, do not make mistakes. However, this does not exclude us as the land surveyors from the need to know the properties and accuracy of electronic measurement of lengths, in engineering surveying when working accurately, even beyond the information of the manufacturer or retailer. The standard deviation of distance D is determined by the manufacturer as

$$\sigma_D = A + B \cdot D \tag{1}$$

According to (Rueger 1990), A is given in millimeters and includes the accuracy of the meter reading, the maximum amplitude (or average) of the short-period cyclical errors in phase-range meters, the maximum (or average) effect of the nonlinear on the length of the dependent errors, and the precision of the prism constant. B is given in ppm (parts per million) and length D is given in kilometers. B for short-range distance meters includes a range of main oscillator drift in the operating temperature range and the maximum error that can be caused by limiting the ppm computation steps. This enumeration is a mix of systematic and random errors. In addition to these phenomena, the meter (operator), the state of the environment, and the accessories used affect the accuracy of measuring lengths. Although planning and determining precision in geodesy usually takes into account only random errors, with systematic errors being adequately suppressed by calibration, comparisons, or measuring procedure, it is not the case of electronic distance meters. This brings uncertainty into the precision thinking and invalidates all common precision calculations where a normal distribution characterizing measurement errors is a prerequisite. The aim of the entire experiment was to determine the size and properties of systematic and random errors of conventional electronic distance meters and whether their properties vary depending on the size of the length. Due to the use of electronic distance meters and required precision, a range of distances of approximately 2 m–50 m was chosen, which corresponds to precise measurements in engineering surveying, and measurements can be made on laboratory interferometers. Longer pillar baselines are not suitable for this purpose, as the point position is stable and the number of points small for the purpose of determining the detailed error map. Based on these hypotheses and for the reasons mentioned above, an experimental procedure for the calibration of electronic distance meters was designed at the Department of Special Geodesy of the Faculty of Civil Engineering CTU in Prague to determine the magnitude of systematic errors, to suppress them and to achieve more accurate measurement results. The procedure is designed for very precise measurements in industry, short lengths (up to 40 m) and closed industrial halls where stable atmospheric conditions are expected.

For the experiment, a new laboratory pillar base with forced centerings and an accuracy of 0.02 mm in absolute lengths was used at the Faculty of Civil Engineering in Prague (Czech Republic) and a laboratory base with an interferometer at the Department of Geodesy, TU Dresden (Germany). Eight different distance meters (20 instruments) were tested and complete tests and verification measurements of the suitability of the proposed calibration procedure were performed on 3 selected.

2 EXPERIMENT

The basic idea of the experiment is to refine EDM measurements of the total station by the knowledge of systematic errors of varying with distances. The magnitude of random errors can be dramatically reduced by a higher number of repetitions according to (Koch, 1999). Moreover, as is well known from practice, random errors are significantly smaller at short distances than stated by the manufacturers. The remaining error of the measured length already contains practically only a systematic component, which can be determined by comparing the specified length with the length determined by a significantly more accurate method.

2.1 *Workflow of the experiment*

The proposed experimental procedure for determining the magnitude of systematic errors at short distances and the introduction of corrections of directly measured lengths for electronic distance meters consists of 9 follow-up steps.

1. Selection of a device with a reference to the length standard to determine absolute lengths.
2. Construction of the laboratory pillar length baseline.
3. Selection and testing of measuring accessories on the laboratory baseline.
4. Determination of the absolute lengths of the laboratory baseline.
5. Determination of absolute error sizes of selected rangefinders and their stability over time.
6. Test the same types of rangefinders and compare the absolute error sizes.
7. Determination of the detailed map of relative errors in 3 range-finders.
8. Determination of the correction function of the three rangefinders from the combination of absolute and relative errors.
9. Experimental verification of the suitability and accuracy of corrections of the rangefinders.

2.2 *Reference instrument*

Total station distance meters measure lengths normally with a 0.1 mm resolution, so it was necessary to select a reference instrument that can determine reference absolute lengths with an accuracy better than 0.05 mm. Another condition was the metrological connection to the national etalon of lengths.

This is why the Leica Absolute Tracker AT401 was used (Fig. 1), which is used to determine the lengths of outdoor pillar calibration bases. The instrument was lent from the Czech

Geodetic, Topographic and Cadastral Research Institute (VÚGTK), which is a state organization responsible for maintaining Czech calibration bases, a state standard for long lengths and issuing calibration sheets for geodetic measuring instruments.

2.3 *Laboratory pillar baseline*

The baseline is located in the geodetic cellar laboratory in the Building C of the Faculty of Civil Engineering of the Czech Technical University in Prague. It consists of 16 concrete pillars in one row (Fig. 2). The height of the pillars is 0.9 m and the square head size is 0.35 m × 0.35 m.

The pillars are 0.9 m–5.0 m apart and the total length of the baseline is 38.6 m. The temperature of the laboratory is around 20°C throughout the year. In 2013, the pillars were fitted with centering plates (Fig. 1), which are lined in one direction with a maximum transverse deviation of 2 mm.

2.4 *Absolute baseline lengths determination*

It was necessary to determine the horizontal distance between the pillars with an accuracy better than 0.05 mm. In July 2013, the base length was determined firstly. The Leica Absolute Tracker AT401 described above was used. For the measurements, 3 standpoints were selected directly on the pillars of the base (pillar 1 and 16 and middle pillar 10) and one eccentric standpoint (1.5 m from the axis of the base near the pillar 6). From each standpoint it was always measured on all pillars of the base. The measurement was performed 1× at two faces. After the adjustment it was calculated from the coordinates of horizontal lengths between the individual pillars, which are characterized by a standard deviation of 0.02 mm. In January and July 2014, control measurements were performed to confirm the overall stability of the base with good result, baseline lengths were not differing more than 0.05 mm (maximal permissible difference with probability 95%).

2.5 *Determining the detailed course of relative errors of selected rangefinders at the base with an interferometer*

The laboratory baseline with an interferometer is highly suited for detecting a detailed course of the systematic errors, but the relative ones are usually determined and must be converted to

Figure 1.   Leica Absolute Tracker AT401 and Leica Red-Ring reflector 1.5″ (RRR).

Figure 2.   Scheme of the laboratory pillar baseline.

109

Instrument	Distance meter type	Unit length	Standard deviation σ_D
Leica TC1202	Phase shift	1,5 m	2 mm + 2 ppm
Leica TS06	Phase shift	1,5 m	1,5 mm + 2 ppm
Leica TC1800	Phase shift	3,0 m	1 mm + 2 ppm
Leica TC2003	Phase shift	3,0 m	1 mm + 1 ppm
Topcon GPT7501	Time of flight	–	2 mm + 2 ppm
Trimble S6 HP	Phase shift	0,37 m	1 mm + 1 ppm
Trimble S8	Phase shift	0,37 m	0,8 mm + 1 ppm
Trimble M3	Time of flight	–	3 mm + 2 ppm

absolute. For testing, a baseline with interferometer was selected at the geodetic laboratory at TU Dresden. The laboratory base consists of a rail on which the cart with prisms is moved.

2.6 Determination of absolute systematic errors

The results from the determination of absolute and relative errors had to be merged. On the data obtained at the base with the interferometer (length and corresponding relative error) an interpolation of the deviations corresponding to the lengths from the laboratory pillar base was applied. From the differences of the corresponding deviations from both experimental procedures the average data shift from the base with the interferometer was calculated, i.e. the relative errors were converted to absolute errors. Since all the deviations are subjected by measurement uncertainty, the weighted average was used. The weighted average calculation is iterative and the individual differences are assigned to the scales using the L1 norm.

2.7 Verification of determined systematic errors

In order to confirm the correctness of the considerations and procedures, experimental measurements were carried out in a laboratory with a pillar base. The test consisted of comparing the lengths (coordinates) determined by the Leica Absolute Tracker AT401, with lengths measured by total stations with corrected lengths.

2.8 Tested instruments overview

Table 1 lists tested devices with important parameters. The aim was to test the pulse and phase distance metres, which have a standard deviation of measured distance of 1 mm–3 mm. For the tests, the instruments at the disposal of the Faculty of Civil Engineering of the Czech Technical University in Prague was used. Measuring principle of the time of flight and phase shift distance meters is fully described in (Rueger, 1990).

3 EXPERIMENT RESULTS

In solving the whole problem, a large number of results have been produced that cannot be presented here, only some interesting results illustrating the main ideas will be shown. Random error measurements of lengths for all total stations can usually be characterized by a standard deviation of approximately 0.3 mm or less, so for their practical suppression it is sufficient to use a mean of 16 measurements. Measurement at two faces is the basic approach in quality measurement, but the differences between them have been tested. In all cases, the difference was less than or equal to 0.1 mm or at the minimum reading step. Systematic errors of measurement of the distances have significant values, which can be seen on two examples in Figures 3 and 4.

Figure 3 shows the Trimble S6 test results, where considerable variability of systematic errors is evident. The shortest wavelength corresponds to the modulated wavelength of a 0.40 meter.

On the other hand, the Trimble M3 (a branded Nikon) is another distance meter with different features, measurement errors are also distance-dependent but significantly less oscillating. These two examples illustrate the diversity of the instruments, and more interesting examples can be found in the article (Braun, 2015). Also interesting is the comparison of devices of the same type and the manufacturer, as an example the absolute errors of the six Topcon GPT-7501 instruments is shown at Fig. 5.

Here we can see that the error pattern is very similar for a given type of device, but there is only a constant shift that could be removed by the quality determination of the prism constant. For other tested series of the same instruments from the other manufacturers, similar conclusions can be drawn. Different behavior is shown only by Leica tested instruments where the measurement errors are small and on the other hand, they have a virtually linear course depending on the distance.

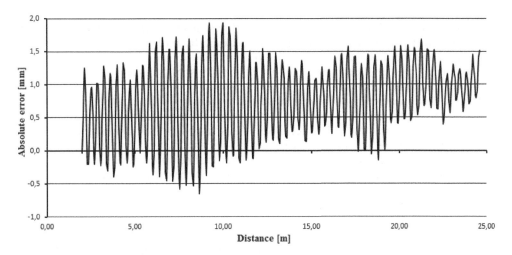

Figure 3. Absolute systematic errors—Trimble S6.

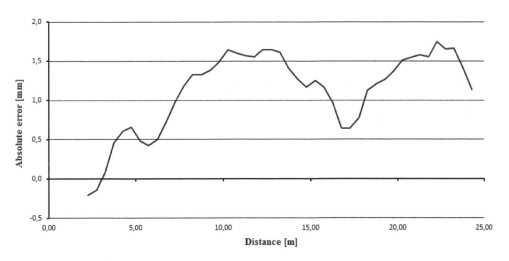

Figure 4. Absolute systematic errors—Trimble M3.

111

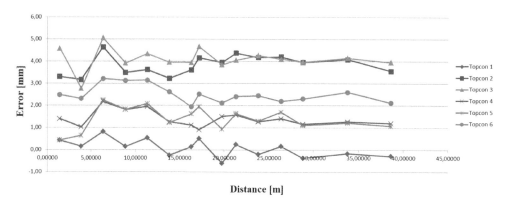

Figure 5.   Absolute systematic errors – 6× Topcon GPT-7501.

4   CONCLUSIONS

The experiment has shown that the hypothesis of systematic errors of electronic distance meters is correct. The usual way of testing the electronic distance meters, where the constant component and the scale is determined, is also not suitable for suppressing systematic errors of the rangefinders for short distance measurements. It is not sufficient, although in some cases it might considerably help.

The verification experiment showed that the application of this knowledge could reduce the errors of the measured distances to standard deviation of approximately 0.5 mm (for short distances, for any tested distance meter). This value is, in our opinion, largely influenced by the used tools and accessories and could be further reduced. Systematic errors continue to be a very dangerous part of measurement errors, and for example geodetic networks of similar lengths, there is a risk that such an error will not be detected in adjustment.

ACKNOWLEDGMENTS

SGS17/067/OHK1/1T/11 Optimization of acquisition and processing of 3D data for purpose of engineering surveying, geodesy in underground spaces and laser scanning.

REFERENCES

Braun, J., Štroner, M., Urban, R. and Dvořáček, F. 2015. Suppression of Systematic Errors of Electronic Distance Meters for Measurement of Short Distances. Sensors—Open Access Journal. 2015, vol. 15, no. 8, p. 19264–19301. ISSN 1424-8220.

Koch, K.R. 1999. Parameter Estimation and Hypothesis Testing in Linear Model, 2nd ed.; Springer Heidelberg: Berlin, Germany.

Rueger, J.M. 1990. Electronic Distance Measurement, 3rd ed.; Springer-Verlag: Berlin Heidelberg, Germany, 1990; pp. 186–221.

Advances and Trends in Geodesy, Cartography and Geoinformatics – Molčíková et al. (Eds)
© 2018 Taylor & Francis Group, London, ISBN 978-1-138-58489-1

Mapping of spoil heap by commercial and home-assembled fixed-wing UAV systems

R. Urban, B. Koska & M. Štroner
Department of Special Geodesy, Faculty of Civil Engineering, CTU in Prague, Czech Republic

ABSTRACT: Technology of Unmanned Aircraft and helicopters (UAV) system is rapidly becoming popular in many research and industry sectors. Due to the large variability and the relatively low purchase price, unmanned systems are largely enforced in geodesy for mapping smaller areas. The article deals with the use of UAV systems for mapping of the spoil heap, which is located next to the surface quarry. Two selected UAV systems were used for mapping. The first system is eBee by company SenseFly. The second system consists of EasyStar II by Multiplex and autopilot 3DR Pixhawk B. The area of spoil heap was measured by both systems in the smallest vegetation period in a very short time range. For the photogrammetry solution, the ground control points were stabilized in the area, and determined by GNSS (method RTK). Both systems were set up similarly (overlapping, resolution) and so it was possible to compare them in output quality, processing speed and economic aspects. The best was measurement by Easystar and then the combination of all flights together from eBee. Thus, it is clear that multiple measuring by UAV can increase the accuracy and reliability of the results. In this case, the disadvantage is the longer processing, but which is, however, completely automatic nowadays. From an economic point of view, it is more advantageous to use freely available aircraft and camera with a fixed focus than the entire commercial system.

1 INTRODUCTION

In these days the unmanned aircraft vehicle (UAV) are currently being used in many sectors of science (Fraštia, 2014), (Blišťan, 2016) and due to their economic aspects (Kršák, 2016), are increasingly available in the commercial sphere. For processing is typically used specialized software for aerial photogrammetry (Bartoš, 2014). Digital photogrammetry methods are based on images from passive sensors. Many UAV platforms equipped with various cameras, differing in their suitability for mapping (Torresan et al. 2017). The accuracy of point clouds or derived products such as DTM can be affected by the camera quality (Thoeni et al. 2014) and cameras with a fixed focal length are considered superior for photogrammetry applications (Shortis et al. 2006). Furthermore, adjacent image overlaps can vary due to interference of external factors such as gusts of wind with the UAV stability (Jensen and Mathews, 2016). Therefore, more robust and stable UAV platforms are likely to provide better results.

The article deals with the assessment of the accuracy of two different UAV systems in the area of spoil heap, where a very diverse terrain can be found. In this type of vegetation the evaluation of point cloud is strongly influenced by a number of systematic errors that arise for various reasons.

2 LOCATION

For experimental measurement the part of Horní Jiřetín spoil heap was chosen (Fig. 1). This spoil heap has a very rugged terrain in the form of hills, cliffs, plains, and water areas, but also diverse stands such as trees, shrubs and tall grass. It is located between Litvínov, Horní Jiřetín and the industrial area of Záluží. The area of this territory is about 410 ha, but only part of this territory was used for the experiment. The spoil heap as an external dump for lignite mining from the mine "Obránců míru" was used. The reclamation took place here from 1969 until 2006. Thanks to the earlier devaluation by landscape mining and subsequent reclamation, nowadays there are animal species that are unparalleled in these countries.

The part of spoil heap (Fig. 2) was equipped with twenty ground control points (blue markings) for measuring with UAV systems. Ground control points was made from white wooden boards 40 × 40 cm with a center black mark of 15 cm radius, which should have the dimensions of 5 × 5 pixels in the image at the intended resolution. These points were determined by Trimble Geo XR GNSS device. Real time kinematic method with a 15-second observation time and by connecting to the network of virtual reference station CZEPOS was used.

Figure 1. Spoil heap.

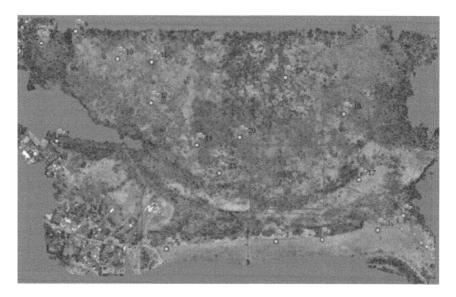

Figure 2. Ground control points (blue markings) and cut-out (red square).

3 UAV SYSTEMS

The site of the spoil heap was measured by two UAV systems in a short span of time during vegetative rest. For both systems a longitudinal overlap of 85%, a transverse overlap of 65% and a resolution of about 3 cm per pixel was set.

Height and speed of the airplane, together with the shutter speed, in both systems control was set by control program depending on the resolution in fact. For both systems, the measuring height was about 100 m above the terrain.

3.1 *SenseFly eBee*

EBee (Fig. 3) from company SenseFly is a classic aircraft with removable wings and a push propeller. It allows an automatic short flight with a flight time of up to 40 minutes. Various cameras for visible and near-infrared radiation are available, and multispectral and thermal cameras are also available. It is a mapping and monitoring system for data acquisition, which is equipped with a complete control unit. For the experiment, the aircraft was equipped with a camera Sony Cyber shot DSC-WX220 with a resolution of 18.2 MPix and a focal length of 4.45 to 44.5 mm.

By eBee system two complete measuring of the spoil heap were realized. Both flights was perpendicular to each other. The first flight had 1004 images (eBee 12) and the second 903 images (eBee 34).

3.2 *Easystar II and 3DR Pixhawk*

EasyStar II is a commercially available motor glider (Fig. 4), which also features a commercially available 3DR Pixhawk autopilot. Autopilot can convert any type of model into an autonomous system. The system also features a Nikon Coolpix A camera with a 16.2 megapixel resolution and a fixed focal length of 28 mm.

By Easystar II glider only one measuring (940 images) of the spoil heap was realized.

3.3 *Analysis of images from UAV*

When comparing data from both systems for the first time, the significant differences in image quality were found. Images from the eBee system included large errors especially at the edges of images where the pixels are blurred. It was probably due to distortion of lens. There

Figure 3.   SenseFly eBee.

Figure 4.   Easystar II.

Figure 5.   Trajectories of flight by system eBee.

is also a difference in flight trajectories where the navigation unit is generally influenced by the height distribution of the interest area. In Fig. 5 are shown camera positions and interlaced curve of two perpendicular flights by the eBee system, where the height differences of the individual lines can be observed. In the case of the height form of the terrain in the direction of the flight rows, the trajectory lines are different in height (the red curve), while perpendicular to the direction of the flight are all trajectories at one height (yellow curve).

4 DATA PROCESSING

Data processing in several specialized software was done. In the Agisoft's PhotoScan, aerial photogrammetry were processed, and the resulting point clouds in the CloudCompare open source program were compared.

4.1 *Agisoft PhotoScan*

Software Photoscan uses epipolar geometry and pixel correlation for processing of aerial photogrammetry. Solving of individual flights was done separately (Easystar, eBee 12, eBee 34), but also flight eBee 1234 (consist of ebee 12 and eBee 34) was created. All of flights were georeferenced by ground control points with a priory accuracy of 0.02 m. The alignment was subsequently completed using accuracy parameter set to 'high' and pair pre-selection to 'disabled'. Accuracy set to 'high' ensured the use of the original image resolution while 'disabled' pair pre-selection ensured the most accurate image matching. The limit for key points (indicating the maximum number of points sampled within each image) was set to 20,000 and for tie points (the number of points used for image matching) to 5,000. A posteriori accuracy of ground control points was evaluated by calculating their coordinate error from the model. 6 points were eliminated from processing for probably degradation (displacement about several meters due to animals or the complete destruction of the point). This devaluation was caused by a time gap between the measuring of the points and the realization of the flights. It was necessary to wait for the appropriate conditions (snow cover, wind speed, etc.). To determine

Table 1. Precision overview of flights.

Flight	Total coordinate error [m]
eBee 12	0.081
eBee 34	0.053
eBee 1234	0.050
Easystar	0.041

Table 2. Point cloud differences.

Flights	Mean difference [m]	Total error [m]	Std. deviation [m]	No. of outliers
Easystar – eBee 12	0.323	0.56	0.46	0
Easystar – eBee 34	0.250	0.38	0.29	0
Easystar – eBee 1234	0.197	0.28	0.19	2
eBee 1234 – eBee 12	0.315	0.62	0.54	0
eBee 1234 – eBee 34	0.085	0.20	0.18	0
eBee 34 – eBee 12	0.365	0.67	0.56	0

the accuracy of the photogrammetric model, the total spatial deviation for all ground control points (the quadratic mean of the total errors of the individual ground control points) were calculated. The precision overview is in Table 1, which shows that the best results are achieved by Easystar, which has the highest quality images. Very good results have been achieved by the eBee 34 flight, where the flight trajectory appropriately copies the chosen terrain.

4.2 *Cloudcompare*

Software CloudCompare was used to compare point cloud. Software calculates difference between two referenced clouds. For best comparison of models the same cut-out (250 × 250 m) in each point cloud was made. The cut-out is suitably located between the ground control points (Fig. 2 – red square). In addition, 100 suitable control points were selected from this cut-out, where the differences between point clouds were calculated. These differences were evaluated by robust method (L1 norm) for finding outliers (Třasák, 2014). Subsequently, the values of the total error of the differences were calculated as a quadratic mean. The total error of the differences contains systematic error and random error. After the reduction of all differences by the mean difference value (potential systematic error), the value of the standard deviation, which already contains only the random error component, was calculated. Table 2 shows the individual parameter of point cloud differences with the number of outliers that were detected by the robust method.

From the results shown in Table 2 it is obvious that almost no outliers have been detected and that the combination with the eBee 12 flight reaches the highest average differences. The smallest systematic error is in the combination of eBee 1234 and eBee 34, which is caused by big correlation of data sets, where the accuracy of the eBee 34 flight significantly determines the accuracy of the eBee 1234.

5 RESULTS

For comparison of individual flights, a simple adjustment (method of least squares) was chosen. The adjustment was based on the six equations of differences between flights shown in Table 2. The equations were determined by the law of accumulation of standard deviations. The standard deviation of the difference between flights in quadrate is given by the sum of the standard deviations of relevant flights in quadrate. Thus, a system of six equations (differences of flights) with four unknowns (the accuracy of individual flights) was created. The system was solved as a linear after a suitable substitution, and in this case the pseudoinversion method based on SVD decomposition (Hampacher, 2015) was used.

The adjustment results are shown in Table 3, which shows that the UAV with best camera has the smallest standard deviation. The partial result was, that suitable combination of perpendicular flights can be achieved good result with a worse camera. Unfortunately, it is also clear that inappropriately selected flight strategies can negatively affect the results achieved.

It is obvious, that flights are interdependent and then result from adjustment is only an estimate of the accuracy. Despite that, it is possible take these results for better planning of measurement of spoil heap by UAV.

Table 3. Standard deviations of flights.

Flights	Standard deviation [m]
eBee 12	0.49
eBee 34	0.23
eBee 1234	0.14
Easystar	0.07

6 CONCLUSION

An experimental measuring of the part of the spoil heap was realized by two unmanned aerial systems. The first system was the commercially available eBee from SenseFly and the other was built from the airplane Easystar II and control unit Pixhawk 3DR. By system eBee, two perpendicular flights and one flight by Easystar were made. The flights were performed in a short time span with the same overlap and resolution parameters. The same ground control points for data processing of every flights were used. Two perpendicular eBee flights were processed together. The smallest deviations at ground control points were achieved by the Easystar, which is due to the quality of the camera. From the resulting point clouds of each flight a cutout was determined. In cutout 100 control points were made and in these points clouds were compared. The best was measurement by Easystar and then the combination of all flights together from eBee. Thus, it is clear that multiple measuring by UAV can increase the accuracy and reliability of the results. In this case, the disadvantage is the longer processing, but which is, however, completely automatic nowadays. From an economic point of view, it is more advantageous to use freely available aircraft and camera with a fixed focus than the entire commercial system. From a practical point of view, this is exactly the opposite, because the complete system guarantees reliability and simplicity. The time-consuming process of capturing images is not very large, so it seems very advantageous to take two sets of data of the same territory if available only worse camera.

ACKNOWLEDGEMENT

This work was supported by the Grant Agency of the Czech Technical University in Prague, grant No. SGS17/067/OHK1/1T/11 and by Czech Science Foundation (project No. 17-17156Y).

REFERENCES

Bartoš K. & Pukanská K. & Sabová J. Overview of Available Open-Source Photogrammetric Software, its Use and Analysis. International Journal for Innovation Education and Research (IJIER). Vol. 2, no. 4 (2014), p. 62–70, ISSN 2201-6740.

Blišťan, P. & Kovanič, Ľ & Zelizňaková, V. & Palková, J. Using UAV photogrammetry to document rock outcrops. Acta Montanistica Slovaca, Volume 21 (2016), number 2, pp. 154–161, ISSN 1335-1788.

Fraštia M. & Marčiš M. & Kopecký M. & Liščák P. & Žilka A. Complex geodetic and photogrammetric monitoring of the Kraľovany rock slide. In Journal of Sustainable Mining. Vol. 13, no. 4 (2014), s. 12–16. ISSN 2300-1364.

Hampacher, M. & Štroner, M. 2015 *Zpracování a analýza měření v inženýrské geodézii.* Praha: CTU Publishing House, p. 336 ISBN 978-80-01-05843-5.

Jensen, J.L., and A.J. Mathews. 2016. "Assessment of image-based point cloud products to generate a bare earth surface and estimate canopy heights in a woodland ecosystem." Remote Sensing, 8: 50. doi:10.3390/rs8010050.

Kršák B. & Blišťan, P. & Pauliková, A. & Puškárová, P. & Kovanič, Ľ. ml. & Palková, J. & Zelizňaková, V. Use of low-cost UAV photogrammetry to analyze the accuracy of a digital elevation model in a case study. Measurement. Vol. 91 (2016), p. 276–287. ISSN 0263-2241.

Shortis, M.R., C.J. Bellman, S. Robson, G.J. Johnston, and G.W. Johnson. 2006. "Stability of zoom and fixed lenses used with digital SLR cameras." International Archives of Photogrammetry and Remote Sensing, 36.

Thoeni, K., A. Giacomini, R. Murtagh, and E. Kniest. 2014. "A comparison of multi-view 3D reconstruction of a rock wall using several cameras and a laser scanner." The International Archives of Photogrammetry, Remote Sensing and Spatial Information Sciences, 40: 573.

Torresan, C., A. Berton, F. Carotenuto, S.F. Di Gennaro, B. Gioli, A. Matese, F. Miglietta, C. Vagnoli, A. Zaldei, and L. Wallace. 2017. "Forestry applications of UAVs in Europe: a review." International Journal of Remote Sensing, 38: 2427–2447. doi:10.1080/01431161.2016.1252477.

Třasák, P. & Štroner, M. Outlier detection efficiency in the high precision geodetic network adjustment. Acta Geodaetica et Geophysica. 2014, 49(2), 161–175. ISSN 2213-5812.

Advances and Trends in Geodesy, Cartography and Geoinformatics – Molčíková et al. (Eds)
© 2018 Taylor & Francis Group, London, ISBN 978-1-138-58489-1

The comparison of selected robust estimation methods for adjustment of measurements in geodetic network

V. Zelizňaková & T. Hurčík

Faculty of Mining, Ecology, Process Control and Geotechnologies, Institute of Geodesy, Cartography and Geographic Information Systems, Technical University of Košice, Košice, Slovakia

ABSTRACT: Each measurement is subject to many disruptive effects that can cause measurement errors. These errors cannot be avoided, but they can be kept within certain limits so that the measurement results are usable. This leads to the implementation of new alternative procedures in a practical adjustment of measurements, in addition to the least square method—robust estimation methods. The presented paper discusses in detail the selected robust estimation method: Huber's robust M-estimation, Hampel's robust M-estimation, robust Tukey's biweight method and Danish method. The paper also compares them with the LSM on the example of adjustment of measurements of the geodetic network established in the underground areas of the Dobšinská Ice Cave.

1   INTRODUCTION

The Gauss-Markov model using the LSM is one of the most commonly used methods for estimation of unknown parameters. However, effective and non-deviated values can be obtained only if the vector of corrections of measured variables has a normal distribution. In the case of hidden gross errors in measurements, the LSM tries to minimise these errors and transfers them to other values.

At the end of the last century, new unconventional procedures that eliminate gross errors from measurements—robust estimation methods (Huber's robust M-estimation, Hampel's robust M-estimation, robust Tukey's biweight method, Welsh's robust M-estimation, Talwar's robust M-estimation, logarithmic M-estimation, or Danish method), have emerged (Huber, 1981; Hampel, 1986; Jäger, 2005; Třasák, 2014).

2   ROBUST ESTIMATION METHODS

In the robust M-estimation (Huber, 1981; Hampel, 1986; Jäger, 2005; Třasák, 2014), the minimised function $_v{}^T{}_v$ is replaced by a more appropriately chose function of corrections $\rho(v)$, referred to as the function of losses:

$$\rho(v) = \min, \tag{1}$$

through which it is possible to create the influence function $\psi(v)$. Then, the following applies:

$$\psi(v) = \frac{\partial \rho(v)}{\partial v}. \tag{2}$$

Consequently, the weight function w(v) is derived from the influence function $\psi(v)$ of the robust M-estimation:

$$w(v) = \frac{\psi(v)}{v}. \tag{3}$$

Table 1. Danish method*.

Function of losses	Influence function	Weight function
$\rho(v)$	$\psi(v) = \begin{cases} v & |v| < c \\ v.e^{-a.|v|^b} & |v| \geq c \end{cases}$	$w(v) = \begin{cases} 1 & |v| < c \\ e^{-a.|v|^b} & |v| \geq c \end{cases}$

*Standard: c = 3. If k = 1 =>a = 0. If k = 2, k = 3 =>a = 0,05. If k > 3 =>b = 3.

Table 2. Huber's robust M-estimation*.

Function of losses	Influence function	Weight function

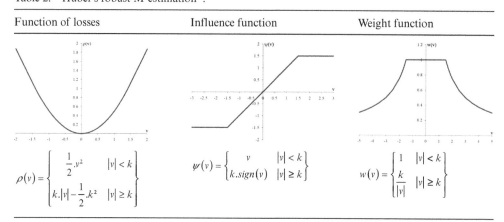

$$\rho(v) = \begin{cases} \dfrac{1}{2}.v^2 & |v| < k \\ k.|v| - \dfrac{1}{2}.k^2 & |v| \geq k \end{cases}$$

$$\psi(v) = \begin{cases} v & |v| < k \\ k.sign(v) & |v| \geq k \end{cases}$$

$$w(v) = \begin{cases} 1 & |v| < k \\ \dfrac{k}{|v|} & |v| \geq k \end{cases}$$

*The value of the damping constant is usually 1,5.

Table 3. Hampel's robust M-estimation*.

Function of losses	Influence function	Weight function

$\rho(v) =$

$$\begin{cases} \dfrac{1}{2}.v^2 & |v| < a \\ a.|v| - \dfrac{1}{2}.v^2 & a \leq |v| < b \\ **\left[1 - \left(\dfrac{c-|v|}{c-b}\right)^2\right] & b \leq |v| < c \\ ** & c \leq |v| \end{cases}$$

$\psi(v) =$

$$\begin{cases} v & |v| < a \\ a.sign(v) & a \leq |v| < b \\ a.\dfrac{c-|v|}{c-b}sign(v) & b \leq |v| < c \\ 0 & c \leq |v| \end{cases}$$

$w(v) =$

$$\begin{cases} 1 & |v| < a \\ \dfrac{a}{|v|} & a \leq |v| < b \\ a.\dfrac{c-|v|}{(c-b).|v|} & b \leq |v| < c \\ 0 & c \leq |v| \end{cases}$$

$** = a.b - \dfrac{1}{2}.a^2 + \dfrac{1}{2}.a.(c-b)$

*Standard: a = 2, b = 4, c = 8.

Table 4.   Tukey's biweight estimation method*.

Function of losses	Influence function	Weight function

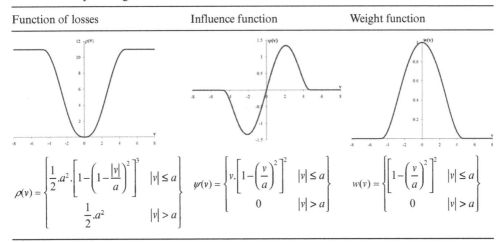

$$\rho(v) = \begin{cases} \dfrac{1}{2}.a^2.\left[1-\left(1-\dfrac{|v|}{a}\right)^2\right]^3 & |v| \le a \\[2ex] \dfrac{1}{2}.a^2 & |v| > a \end{cases}$$

$$\psi(v) = \begin{cases} v.\left[1-\left(\dfrac{v}{a}\right)^2\right]^2 & |v| \le a \\[2ex] 0 & |v| > a \end{cases}$$

$$w(v) = \begin{cases} \left[1-\left(\dfrac{v}{a}\right)^2\right]^2 & |v| \le a \\[2ex] 0 & |v| > a \end{cases}$$

*Standard: a = 4,685.

Huber's robust M-estimation (Table 2), Hampel's robust M-estimation (Table 3), robust Tukey's biweight method (Table 4), Welsh's robust M-estimation, Talwar's robust M-estimation, logarithmic M-estimation, or Danish method (Table 1) are the most commonly used M-estimation methods (Huber, 1981; Hampel, 1986; Jäger, 2005; Zelizňaková, 2014).

The basic idea of these methods is the condition that large residues indicate less accurate measurements and vice versa. Therefore, after the conventional estimation of parameters by the LSM in the Gauss-Markov model, a priori weights are replaced by new ones that are a function of residues. Subsequently, a new estimation is made, resulting in new residues from which new weights are calculated again. This process of estimation and modification of weights is repeated until convergence is achieved.

Parameters of the last step of iteration are considered as best and final estimates and no measurement is excluded.

3   DOBŠINSKÁ ICE CAVE

The geodetic network (Zelizňaková, 2014) established in the underground areas of the Dobšinská Ice Cave served as the basis for this work. Dobšinská Ice Cave (Figure 1) as a part of the Stratená Cave System was formed in the Middle Triassic pale Steinalm and Wetterstein limestones of the Stratená Nappe, along the tectonic faults and interbed surfaces. The cave reaches the length of 1,491 m and vertical span of 112 m.

The cave is not unknown to surveyors. Several geodetic measurements were carried out there since its discovery by Mr. Ruffíny, Lang and Mega in June 1870 to the present. However, some of the initially monumented survey control points were damaged and lost due to disintegration and impact of water and ice (most recently in 1996). Additional points were destroyed during the change of show paths in the interior and change of pavement at the entrance to the cave. For this reason, 5 survey control points were newly monumented in the entrance area of the Dobšinská Ice Cave (5001, 2003, 5004, 8001, 8002; points are monumented by survey marking nails. In the underground areas of the cave, 7 survey control points were newly monumented (5005, 5006, 5007, 5008, 5009, 5010, 5011; points are monumented in the ceiling using nails specially adapted to plumb bob, surveying prism and centring of an instrument below the point). From previous geodetic measurements, monumentation of 4 survey control points was taken (7013, 7018, 7020, 7021), whose coordinates were redefined in the coordinate system S-JTSK (Datum of Uniform Trigonometric Cadastral Network). The geodetic network (Figure 2) consists of 11 points. 5 points form a one-sided connected and oriented traverse. Points 8001 and 8002 serve as witness points in case of damage or destruction of the point 5001.

Figure 1. Ice filling of the Dobšinská Ice Cave.

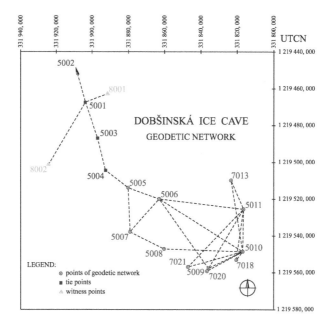

Figure 2. Geodetic point field of the Dobšinská Ice Cave.

4 THE EXPERIMENTAL COMPARISON OF ROBUST ESTIMATION METHODS IN THE ADJUSTMENT OF GEODETIC NETWORK

For the comparison of results, the estimation of parameters of the 1st order of the geodetic network was also realised by the Least Squares Method (Ghilani, 2006):

$$\mathbf{v} = \mathbf{A}.\mathbf{d\hat{C}} - \mathbf{dl},$$
$$0 = \mathbf{G}^T.\mathbf{d\hat{C}},$$
$$\Sigma_l = \sigma_0^2.\mathbf{Q}_l,$$
(4)

where **v** is a vector of corrections of measured variables,
  A is a matrix of design,
  $\mathbf{d\hat{C}}$ is a vector of the adjusted coordinate complements,
  dl is a vector of the measured variables complements,
  G is a datum matrix,
  Σ_l is a covariance matrix of measured variables,
  σ_0^2 is a unit a priori variance factor,
  $\mathbf{Q}_l$ is a cofactor matrix of measured variables.

Table 5. Coordinates of points of the geodetic network.

| | LSM | | HUBER, HAMPEL, TUKEY, DANISH M. | |
| | X | Y | X | Y |
Point	m	m	m	m
5005	1219513,959	331880,840	...,959	...,840
5006	1219520,031	331863,605	...,031	...,605
5007	1219537,898	331879,790	...,897	...,790
5008	1219547,233	331860,951	...,232	...,952
5009	1219557,762	331836,520	...,761	...,521
5010	1219548,690	331817,615	...,689	...,615
5011	1219525,537	331817,182	...,537	...,183
7013	1219509,954	331823,952	...,954	...,952
7018	1219553,159	331821,131	...,158	...,132
7020	1219559,308	331837,318	...,307	...,318
7021	1219557,039	331847,769	...,038	...,769

Table 6. Accuracy characteristics of the points of the geodetic network.

| | LSM | | | | HUBER | | HAMPEL | | TUKEY | | DANISH M. | |
| | s_X | s_Y | s_{XY} | s_P | s_{XY} | s_P | s_{XY} | s_P | s_{XY} | s_P | s_{XY} | s_P |
Point	mm	mm	mm	mm	mm	mm	mm	mm	mm	mm	mm	mm
5005	0,00	0,00	0,00	0,00	0,00	0,00	0,00	0,00	0,00	0,00	0,00	0,00
5006	0,22	0,63	0,47	0,66	0,67	0,95	0,69	0,97	0,60	0,85	0,69	0,98
5007	0,61	0,96	0,80	1,14	1,12	1,58	1,14	1,61	1,01	1,42	1,14	1,62
5008	0,91	1,34	1,15	1,62	1,48	2,09	1,50	2,13	1,32	1,87	1,51	2,14
5009	1,63	1,89	1,77	2,50	2,04	2,88	2,06	2,91	1,80	2,55	2,06	2,92
5010	2,20	1,40	1,84	2,60	2,19	3,10	2,22	3,14	1,93	2,72	2,23	3,15
5011	2,22	0,80	1,67	2,36	1,99	2,81	2,01	2,85	1,75	2,48	2,02	2,86
7013	2,07	1,41	1,77	2,50	2,15	3,04	2,18	3,09	1,92	2,71	2,19	3,10
7018	2,08	1,64	1,87	2,65	2,25	3,19	2,28	3,23	1,99	2,81	2,29	3,24
7020	1,63	2,00	1,82	2,58	2,18	3,08	2,20	3,12	1,92	2,71	2,21	3,13
7021	1,58	1,90	1,75	2,47	2,09	2,95	2,11	2,99	1,85	2,62	2,12	3,00
average			**1,36**	**1,92**	1,65	2,33	1,67	2,37	1,46	2,07	**1,68**	**2,38**

Legend: s_X, s_Y – standard deviations of coordinates,
s_{XY} – mean coordinate error of point,
s_P – mean positional error of point.

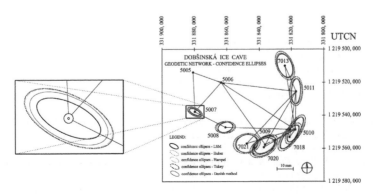

Figure 3. Confidence ellipses.

Since the number of measured variables (n = 44; 19 measured lengths and 25 measured directions) is greater than the number of determined parameters (k = 22; 11 points of the geodetic network, two coordinates are determined for each point of the network), the network is sufficiently determined, i.e. sufficient number of redundant measurements is ensured.

The Pope's τ-test was used to test the corrections of measured variables for the detection of outliers:

$$T_i = \frac{|v_i|}{s_0 \cdot \sqrt{q_{v_i}}} \approx \tau, \tag{5}$$

where q_{v_i} is a cofactor of corrections of measured variables,
  s_0 is a posteriori variance factor,
  τ is a critical value of test.

The test detected one measurement loaded by gross error – $d_{5006,5009}$. The outlier was excluded from the set of measurements, and the new adjustment with the reduced vector of measured variables was performed. In the second cycle of adjustment, the localisation test did not confirm the presence of outliers.

The following robust M-estimations were used in the next adjustment method of this geodetic network: Danish method, Huber's robust M-estimation, Hampel's robust M-estimation, robust Tukey's biweight method. The results of adjustment are summarised in Table 5 and Table 6.

Confidence ellipses (Figure 3) indicate the propagation of errors away from points 5005 and 5006 of the network.

5   CONCLUSIONS

Geodetic measurements of the ice filling in the Dobšinská Ice Cave realised so far indicate that the filling of the cave is not static but it dynamically responds to climatic and hydrological changes. In order to protect this natural monument, it is important to realise all measurements at the highest level of accuracy. A well-established and monumented geodetic network is a necessary prerequisite. The original network established in the underground areas of the cave was suitable extended by 7 newly monumented points. The objective of this paper was to point out alternative estimation methods, therefore, to compare results, the adjustment of the geodetic network is realised by the LSM and selected robust estimation methods. The maximum difference between the coordinate of the point determined by LSM and robust estimation method is 1 mm. Almost identical results were obtained by adjustment of measurements. Therefore, exclusion of outliers or reduction of the impact of these measurements on the result of adjustment requires an individual approach.

REFERENCES

Ghilani, Ch. D., Wolf, P.R. 2006. Adjustment Computations: Spatial data analysis. Fourth edition. New Jersey: John Wiley & Sons, Inc. 610 pp. ISBN 10 0-471-69728-2.
Hampel, F.R., Ronchetti, E.M., Rousseeuw, P.J., Stahel, W.A. 1986. Robust Statistics: The Approach Based on Influence Functions. New Jersey: John Willey & Sons, Inc. 502 pp. ISBN 978-0471735779.
Huber, P.J. 1981. Robust statistics. Willey & Sons, New Jersey: John Willey & Sons, Inc. 308 pp. ISBN 0-471-65072-2.
Jäger, R., Müller, T., Saler, H., Schwäble, R. 2005. Klassische und robuste Ausgleichungsverfahren. Ein Leitfaden für Ausbildung und Praxis von Geodäten und Geoinformatikern. Heidelberg: Herbert Wichmann Verlag. 340 pp. ISBN 978-3-87907-370-2.
Třasák, P., Štroner, M. 2014. Outlier detection efficiency in the high precision geodetic network adjustment. *Acta Geodaetica et Geophysica.* Volume 49, Issue 2, P. 161–175.
Zelizňaková, V. 2014. Empirical verification of causal relations between robustness analysis of geodetic networks and robust M-estimations of their parameters. PhD. Thesis. Košice, 152 pp.

Part B: Geodetic control and geodynamics

Advances and Trends in Geodesy, Cartography and Geoinformatics – Molčíková et al. (Eds)
© 2018 Taylor & Francis Group, London, ISBN 978-1-138-58489-1

Is Galileo ready for precise geodetic applications?

Ľ. Gerhátová, M. Mináriková & M. Ferianc
Department of Theoretical Geodesy, Faculty of Civil Engineering, Slovak University of Technology
Bratislava, Slovak Republic

ABSTRACT: This paper introduce the possibilities offered by the European navigation satellite system Galileo. Galileo observations can be used for data processing separately, or can be combined with Global Navigation Satellite Systems (GNSS). ITMS permanent stations network has been extended by some other permanent stations from the territory of Slovakia and the surrounding areas. Observations of all GNSS are available in RINEX3, at selected permanent stations. Precise Point Positioning (PPP) and Relative Positioning (RP) were realized using Bernese GNSS software. Different ambiguity resolution strategies were applied. In both approaches (PPP and RP) was realized multi-GNSS data processing with combination of GPS, GLONASS and Galileo and individual solutions. Coordinates comparison of each daily solution from 1 March to 30 April 2017 in consideration of the selected reference position of permanent station was performed and transformed to the local horizontal system (n, e, u). Based on the results of this study Galileo accuracy is not yet at the level of fully operational GPS and GLONASS systems and results also confirmed possibility of Galileo-only positioning in geodetic applications.

1 INTRODUCTION

Global Navigation Satellite Systems (GNSS) and their observations provide, at various levels of accuracy, the ability to determine sites position on Earth surface in real-time as well as post-processing. In the case of continuous measurements, we can obtain short-term and long-term information about time variability of point position.

At present, two GNSSs are fully implemented: the US NAVSTAR GPS (Navigation System with Time and Ranging Global Positioning System) and the Russian GLONASS (Globaľnaya Navigatsionnaya Sputnikowaya Sistema), which have been in the operational phase for nearly 40 years. They provide positioning at the level of a few meters navigation accuracy, sub-metric position accuracy when used in geographic information systems, position accuracy of few centimeters to millimeters in geodetic applications. GNSS provides real-time, near-real-time and post-processing positioning. GNSS is also used to determine time systems. The disadvantage of these systems is that they are primarily military and their activities could be limited for civilian use in the case of international political conflicts (Hoffmann-Wellenhof et al. 2008).

The first idea of the primary civil GNSS came from the European Union (EU) in 1990. Since 1999, the EU, in cooperation with the European Space Agency (ESA), has begun to develop the first civilian European global navigation satellite system Galileo. The system is currently in building up phase and it is expected to be finished in 2020. The system is designed to be compatible with all existing GNSSs and capable of working with US GPS and Russian GLONASS (Hoffmann-Wellenhof et al. 2008).

At present, majority of processing softwares allow to use Galileo satellite observations. However, not much attention is aimed at Galileo data processing. The purpose of this paper is to focus on current possibilities of Galileo from the point of view of basic methods of GNSS positioning—both individually as well as in combination with fully operational systems GPS and GLONASS.

The development of the system was not at all easy in the beginning. The project was funded from the EU as well as from private sources within Public—Private Partnership. In 2003, international cooperation started with the aim of cooperation of Galileo with other GNSSs and building international infrastructure to operate the system. Among current partners of Galileo there are e.g. the USA, Russian Federation, China, Israel, Ukraine, Morocco, South Korea, Norway, Canada, India, Australia, Brazil, Chile and Malaysia (Hoffmann-Wellenhof et al. 2008).

The development program of Galileo was divided into three basic phases: Galileo System Test Bed (GSTB), In-Orbit Validation (IOV) and Full Operational Capability (FOC). In the first testing phase (GSTB), two test satellites were launched—GIOVE-A in 2005 and GIOVE-B in 2008 (ESA 2017). Their task was to characterize radiation and magnetic field in the planned orbits, to test the performance of atomic clock, influence of radiation on digital technology in the satellite, transmission of experimental signals, determination of Galileo System Time (GST) and testing the accuracy of signal generated by satellites (Hoffmann-Wellenhof et al. 2008). Their activity was terminated in 2012. IOV was the second phase when four satellites (IOV-1 to IOV-4) were launched. They were deployed to the orbit in pairs by Soyuz-STB launcher from European Space Port in Kourou, French Guiana. Their task was to check the conception based on synchronized utilization of four satellites and limited ground system configuration. They are a part of final constellation of Galileo satellites. FOC is the last phase of the Galileo program. Its task is to finalize building of space segment and international ground infrastructure. Fourteen satellites were launched from Kourou by the end of 2016 (FOC-1 to FOC-14).

From the point of view of internal structure, the European global navigation satellite system Galileo consists of three basic components (ESA 2017). The global component consists of Galileo space segment and Galileo ground segment and is the core of the whole system. The space segment after full operational capability, planned by 2020, will comprise 30 satellites (24 in full service and 6 in-orbit spares), homogenously positioned in three Medium Earth Orbits (MEO). The position is marked as Walker 24/3/1 configuration (European GNSS Agency 2017). The ground segment consists of Ground Control Segment (GCS) and Ground Mission Segment (GMS). GCS is responsible for management of satellites. The task of GMS is to continuously monitor and control the satellite system, check and store the data, determine GST in accor-dance with International Atomic Time (TAI) and the Galileo terrestrial reference frame (GTRF), periodically create and uplink navigation message for each satellite and provide navigation services. Regional components provide independent information about the integrity of satellite signals by External Region Integrity Systems (ERIS) and these will be transmitted by special authorized channels. Local components are deployed for enhancing the performance of Galileo locally (e.g. airports, maritime harbor). These will enable higher performance such as the delivery of navigation signal in areas where the satellite signals cannot be received (e.g. inside buildings, tunnels) (ESA 2017). User receivers and terminals provide basic applications such as localization, navigation, research, precise time etc.

Every Galileo satellite transmits 10 navigation signals in four frequencies (E1, E5a, E5b and E6) in L-band of electromagnetic waves. Signals are right-hand circularly polarized. Six signals on carrier frequencies E1, E5a and E5b will be available for all users (Hein et al. 2002). Two signals on E6 frequency will be encrypted and reserved for Commercial Service users. The last two signals on E1 and E6 frequency will be encrypted and reserved only for Public Regulated Service users. When full constellation of satellites will be achieved, Galileo would provide four services for navigation, positioning and precise timing worldwide (European GNSS Agency 2017): Open Service (OS), Public Regulated Service (PRS), Search and Rescue Service (SAR), Commercial Service (CS). On 15 December 2016 the European Commission announced officially the launch of first three free services (OS, PRS and SAR) (Gibbons 2016). They use exclusively Galileo signals, thus they operate independently of other navigation satellite systems.

3 METHODS OF GNSS POINT POSITIONING

Two basic methods are used for positioning of points on the Earth surface by GNSS observations: point positioning (PP) and relative positioning (RP) and their modifications (Hoffmann-Wellenhof et al. 2008).

Processing of pseudoranges from code measurements in undifferenced form is used for PP in geocentric reference system. It is based on broadcast ephemerides and models of GNSS satellite clock contained in the navigation message. The result of the processing are geocentric coordinates of the point independently from other receivers and reference stations. The disadvantage is low accuracy influenced by the uncertainty of satellite position and clock parameters as well as systematic influences on measurements. Standard deviation of PP is about 2–5 m. The decrease of the influence of systematic effects on PP is possible be means of differential methods (differential GNSS – DGNSS) – application of differential corrections determined in the set of reference stations. The accuracy is this way increased to the level of approximately 1 m.

RP uses measuring carrier phase (or carrier phase differences) performed simultaneously on minimally two points where one is known (reference) and we determine the position of the other one (in accordance with the reference point). We use differenced phase measurements thanks to which most systematic influences are eliminated or reduced. This way it is possible to determine coordinates of the point in accordance with the reference point with several mm accuracy. The disadvantage is that the behavior of the reference point affects the estimated one.

Precise Point Positioning (PPP) is a modification of PP and it achieves accuracy comparable to RP. PPP method uses undifferenced code and phase measurements together with precise orbits and clock parameters of GNSS satellites (Kouba, J. & Héroux, P. 2001). It is necessary to model all systematic influences in processing because they fully influence data processing and interpretation of gained results. The request for correct modelling of all systematic influences is especially important in combination of undifferenced observations of more GNSSs (it is necessary to model relations between individual satellite systems).

4 MULTI-GNSS DATA PROCESSING OF PERMANENT STATIONS NETWORK

In this paper, we present results from data processing of permanent stations network (Figure 1, Table 1) from the project of National Center for Diagnosing the Earth Surface Deformations in Slovakia (ITMS) and the surrounding areas (Hungary – PEN2, BUTE, Poland – ZYWI, USDL).

There are nine geodynamic stations in this project. Existing stations monuments are concrete pillars and newly built stations have special deep-drilled braced monuments for the

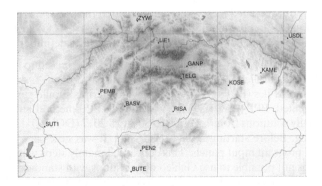

Figure 1. Network of permanent stations used for multi-GNSS data processing.

Table 1. List of selected permanent stations used for multi-GNSS data processing.

Station	GNSS Receiver	GNSS Antenna	Sites	Network
BASV	Trimble NetR9	TRM55971.00	Banská Štiavnica	SK*POS*®, ITMS
BUTE	LEICA GR25	LEIAR25.R4	Budapest, HU	EPN
GANP	Trimble NetR9	TRM59800.00	Gánovce	SK*POS*®, EPN, IGS, ITMS
KAME	Trimble NetR9	TRM59800.00	Kamenica n. Cir.	SK*POS*®, ITMS
KOSE	Trimble NetR9	TRM59800.00	Košice	SK*POS*®
LIE1	Trimble NetR9	TRM59800.00	Liesek	SK*POS*®, ITMS
PEMB	Trimble NetR9	TRM59800.00	Partizánske	SK*POS*®, ITMS
PEN2	LeicaGRX1200+GNSS	LEIAR25.R4	Penc, HU	SK*POS*®, EPN
RISA	Trimble NetR9	TRM55971.00	Rimavská Sobota	SK*POS*®, ITMS
SUT1	Trimble NetR9	TRM59800.00	Bratislava	–
TELG	Trimble NetR9	TRM59800.00	Telgárt	SK*POS*®, ITMS
USDL	Trimble NetR9	TRM59900.00	Ustrzyki Dolne, PL	EPN
ZYWI	Trimble NetR9	TRM59900.00	Zywiec, PL	SK*POS*®, EPN

research of geokinematics and geodynamics. Since the beginning of the project, the equipment on these stations at least allows dual frequency geodetic-type GNSS observations. Data from permanent stations are processed by Bernese GNSS software in the current Version 5.2 by means of automated Bernese Processing Engine (BPE) tool (Dach et al. 2015). Data multi-GNSS processing is conducted according to recommendations and standards in guidelines for EPN Analysis Centres (EPN 2017). Routine data processing of the network was realized in ITRF2008. The network reprocessing started with the publishing of new realization ITRF2014 using observations since 1 February 2017. We also included Galileo satellite measurements at selected permanent stations in the data processing since 1 March 2017.

Data processing consisted of several parts. Multi-GNSS observations of NAVSTAR GPS, GLONASS and Galileo are available in RINEX format, version 3, with 30 s data sampling interval. It is necessary to create a new campaign before processing and prepare a set of project related files (Dach et al. 2015):

– Precise GNSS satellite orbits, clock corrections and Earth rotation parameter (ERP) information – CODE MGEX (Center for Orbit Determination in Europe, The Multi-GNSS experiment) products (marked with COM),
– Ionosphere models for ambiguity resolution and Higher-order ionosphere corrections,
– Differential code biases (DCB) – CODE MGEX products,
– coefficients of the Vienna mapping function (VMF1),
– preparing general files: Antenna phase center variations (PCV.I14), Satellite information file (SATELLIT.I14), Receiver information file (RECEIVER.), Observation type selection file (OBS.SEL – select of observation type for two Galileo frequencies: E1 and E5a),
– preparing set of project related files: Station information file (*.STA), Tectonic plate assignment file (*.PLD), Ocean tidal loading table (model FES2004.BLQ) and Atmospheric tidal loading table (model Ray & Ponte 2003, *.ATL file), Baselines definition file (*.BSL).

Data processing was realized by PPP and RP. The BPE may be used to perform all possible tasks, sequentially or in parallel (Dach et al. 2015). The BPE processing tasks are defined by the user in Process Control Files (PCFs). PCF defines which scripts should run and in what order they should be executed. PPP_BAS.PCF file was used in PPP method and RNX2SNX.PCF was used in RP. Daily solutions from 1 March to 30 April 2017 (DOY 60–120) were the subject of data reprocessing. Before starting a data processing, it was necessary to set or check the original settings of program input panels. This concerned mainly steps where Galileo observations were used (it is not possible to use Galileo observations with standard software settings).

PPP processing consists of few basic steps: copying a set of project files into a campaign, conversion of ERP files, precise ephemerides and satellite clock files from CODE standard format into the internal Bernese format, selection of observation type for two Galileo frequencies in OBS.SEL file, detection of cycle slips, synchronization of receiver clock, and realization of PPP procedure of each station. The results of data processing are geocentric coordinates of GNSS stations with centimeter level accuracy, troposphere and receiver clock parameters. Final coordinates of stations related to IGS14 at the epoch of the observations (Dach et al. 2015).

RP was realized as a double-difference network processing. Further steps in the data processing were following: forming of baselines with BASV reference station, preprocessing and screening of phase data with cycle slips detection, computing a first network solution without resolved ambiguities. An advance ambiguity resolutions schemes were used for ambiguity solving. Baselines with the length up to 200 km were solved by phase-based wide-lane (L5) ambiguity resolution which is based on linear combination of phase data. Code-based wide-lane (WL) ambiguity resolution is used in baselines 200–6000 km long. It is based on linear combination of Melbourne-Wübbena. Consequently, quasi-ionosphere-free (QIF) ambiguity resolution was processed with all data. The resolved ambiguities may be introduced as known into the final network solution. The outputs of the network solution are geocentric coordinates, troposphere estimates, normal equation files with coordinates and troposphere parameters of all stations and only the coordinate parameters in SINEX format (Dach et al. 2015).

5 CONCLUSIONS AND DISCUSSION

In both approaches (PPP and RP) was realized multi-GNSS data processing with combination of GPS+GLONASS (GR) as a reference solution. It was followed by the combination of GPS+GLONASS+Galileo (GRE) and individual solutions of GPS (G), GLONASS (R), Galileo (E). Mean value of each coordinate component in reference solution was calculated. Next step was a calculation of residuals of individual daily solutions towards reference mean value. For most practical applications and interpretations local horizontal coordinates (n, e, u) are preferred. In the next step we transform these residuals to the local horizontal coordinate system. The graph shows these residuals on RISA permanent station in north-south (n) and up (u) component (east-west (e) component is very similar to n component). Individual solutions have been shifted by multiples of 0.02 m from the reference GR solution for clarity. Solution E has a different scale for larger variability of values. The resulting solutions on other permanent stations are identical (except PEN2 and BUTE).

For the PPP method (solutions GR, GRE, G, R), most of the resulting values vary across ± 7 mm in horizontal components and ± 15 mm in the vertical component (Figure 2), which is consistent with the stated accuracy of PPP.

The resulting solutions of RP show greater stability of daily solutions. Values vary between ± 5 mm in horizontal components and ± 10 mm in the vertical component (Figure 3), which is consistent with the accuracy of the static method. For Galileo-only solutions, accuracy is

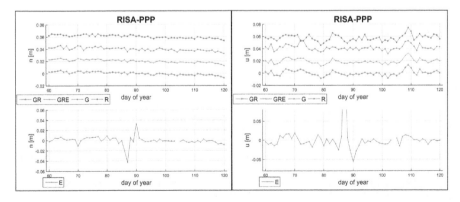

Figure 2. Residuals in north-south (n) and up (u) component from PPP data processing (example of RISA permanent station) for all analyzed solutions.

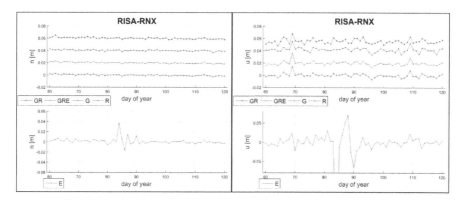

Figure 3. Residuals in north-south (n) and up (u) component from RP data processing (example of RISA permanent station) for all analyzed solutions.

limi-ted by the number of satellites, position and satellite clock correction accuracy, ambigui-ties resolutions, equipment and some other factors. This causes variability of the solution ± 15 mm in horizontal and ± 30 mm in the vertical component. Within the processing interval, we were unable to process the 3–5 days with data unavailability. Special case is permanent station PEN2, where gained results for Galileo solution are significantly worse. Approximately 30% results are in the range ± 120 mm. More outlier values are also observed at BUTE. The cause may be the use of another type of receiver or receiver firmware version.

Based on the daily results of all solutions over two months, it can be stated that Galileo accuracy is not yet at the level of fully operational **GPS** and **GLONASS** systems, but con-tributes to a more stable multi-GNSS combined solution. The results also confirmed that it is possible Ga-lileo-only positioning in geodetic applications. It should be noted that for a comprehensive analysis and a more objective evaluation of the Galileo-only or multi-GNSS solutions with Ga-lileo, it would be necessary to process GNSS observations from a longer time interval and on a larger number of permanent stations, respectively in other areas. It is assumed that the accuracy of Galileo will increase with gradual completion of the system.

ACKNOWLEDGEMENT

This work was supported by the Grants No. 1/0682/16 of the Grant Agency of Slovak Repub-lic VEGA. This paper is result of implementation of projects: National Centre for Diagnos-ing the Earth's Surface Deformations in Slovakia, ITMS 26220220108.

REFERENCES

Dach, R. & Lutz, S. & Walser, P. & Fridez, P. 2015. *Bernese GNSS Software Version 5.2* [online], Bern: University of Bern, Astronomical Institute. Available: <http://www.bernese.unibe.ch>.
EPN – EUREF Permanent GNSS Network 2017 [online]. Available: <http://epncb.oma.be>.
ESA 2017. Galileo System [online]. Available: <http://www.esa.int/ESA>.
European GNSS Agency 2017 [online]. Available: <https://www.gsc-europa.eu>.
Gibbons, G. 2016. European Commission Declares Galileo Initial Services Available for Use. *In Inside GNSS* [online]. vol. 12, no. 1. Available: <http://www.insidegnss.com /node/5268>.
Hein, G.W. & Godet, J. & Issler, J-L. & Martin, J-Ch. & Erhard, F. & Lucas-Rodriguez, R. & Pratt, T. 2002. Status of Galileo Frequency and Signal Design. Available: <http://www.ion.org/publications/>.
Hoffmann-Wellenhof, B. & Lichtenegger, H. & Wasle, E. 2008. *Global Navigation Satellite Systems. GPS, GLONASS, Galileo & more.* Wien – New York: Springer.
Kouba, J. & Héroux, P. 2001. Precise Point Positioning Using IGS Orbits and Clocks Products. *GPS Solutions*, 5 (2), pp. 12–28.

Advances and Trends in Geodesy, Cartography and Geoinformatics – Molčíková et al. (Eds)
© *2018 Taylor & Francis Group, London, ISBN 978-1-138-58489-1*

Application of microgravity for searching of cavities in historical sites

J. Chromčák & J. Ižvoltová
Department of Geodesy, Faculty of Civil Engineering, University of Žilina, Žilina, Slovakia

M. Grinč
INSET s.r.o., Žilina, Slovakia

ABSTRACT: The research in geophysical methods increase the possibility of use in other, never reviewed areas. A construction of the relative gravimeters makes the use method of microgravity possible for the searching of cavities and free spaces in subsoil, not only in opened areas but in urban areas as well. Microgravity method was used for the research in church of Saint Mary in Kláštor pod Znievom village before. This church was chosen after discussion with local historians, they were presuming that the church is that kind of sculpture, where the cavity could be found. After necessary measurements, calculations and creating of Complete Bouguer anomaly map, there is a possibility to define the existence and the position of the cavity. On the other hand, one method is not enough, so there is a necessity to use other method. The method of GPR research shows as a good one, it is not only fast but it is reliable in the same time. As the measurement showed, both of these methods could be used in urban areas. It is good to focus on right calculation of terrain corrections. With the big influence of the walls gravity there can easy come to a mistake of bad interpretation due to Simple Bouguer anomaly. Application of these methods was also used in other historical site, church in Banská Belá. A result of this research supports the correctness of this combination and showed us the same final results if not even better.

1 INTRODUCTION

Hand in hand with the microgravity development, there was an interest to find application of this method to be used. An expansion of the application was also supported by finding of relative gravimeters.

Microgravimetric and gravity gradient surveying techniques are applicable to the detection and delineation of shallow subsurface cavities and tunnels [1]. Geophysical prospection as microgravity is suitable for the detection of the unknown underground voids ever in complex urban settings as well as in historical structures [2, 3]. The best way of the measurements' presentation is construction of Bouguer gravity map. Microgravity surveys seek to detect areas of contrasting or anomalous density by collecting surface measurement of the Earth's gravitational field.

One of many usage possibilities of Bouguer gravity map is finding small underground cavities and free spaces. The finding can be possible after correct measurement method and also after right way of calculation. The unwritten rule is that it is necessary to observe at least three points above an anomaly to define it. Therefore is necessary to choose good density of raster, in witch is place measured. For this measurement are used relative gravimeters. A gravimeter is a high precise instrument used just for measuring the local gravitational field of the Earth [5].

The measured area is nearly always limited by conditions of site, in witch is the measurement done. There is a big importance of cites walls gravity influence elimination. That is the reason for correct documentation of walls shape.

The final Bouguer gravity map is created from Bouguer anomalies, which are calculated on each and every point. The calculation is resulting from formula (1).

$$\Delta g_B = g_{obs} - g_\lambda + \delta g_F - \delta g_B + \delta g_T \qquad (1)$$

where Δg_B is the Bouguer anomaly, g_{obs} is the observed gravity, g_λ is the correction for latitude, δg_F is the well-known free-air correction, δg_B is the Bouguer correction that allows for the gravitational attraction of the rocks between the observed point and sea level and δg_T is a terrain correction [6].

The observed gravity can be measured by relative or by absolute gravimeter. The difference between results is in next processing. In the microgravity research are always used relative gravimeters because of their speed and physical factors, interesting trough measurement (weight, size...). The precision of every relative gravity meter is good enough for this type of activity. Therefore it is not necessary to delete the influence of ellipsoid gravity. But there is necessary to eliminate the mistake caused by gravity meters' observation mode.

As it is possible to see at a Figure 1, there is an example of the instruments' drift processed by DbGrav software, created on Kiel University. On the left side of Figure is the scale of units in mGal. The Figure shows us the drift of the instrument that the influence on the measured data. To reduce or eliminate this influence, it is necessary to measure on a "starting" point, where the gravity has to be still the same, and the only changing element is this instrument drift.

The process of measurement is the same reason, why it is not necessary to eliminate the correction for latitude. The influence of the latitude is in the points of measured area same, that we can consider it as a constant.

The free-air correction is also known as Faye correction. After reduction from measured gravity we become the free-air gravity anomaly or simply called free-air anomaly. This correction is calculated from Newton's Law, as a rate of change of gravity with distance, for Earth this correction is

$$\delta g_F = 0,3086h \qquad (2)$$

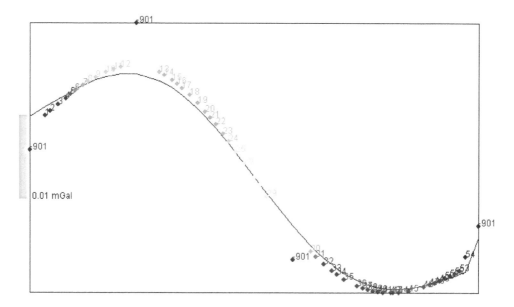

Figure 1.   The process of Gravity Scintrex CG-5 autograv gravity meter.

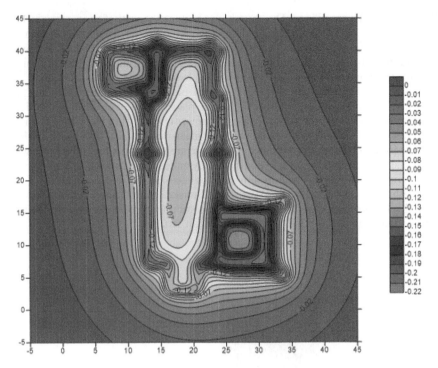

Figure 2.   The values of church walls influence (mGal).

where h is thickness above the reference ellipsoid. If we are doing measurement just in small area, there is not necessary to do it in reference vertical datum at all. It is enough to do it in relative vertical datum. On the other hand, the free-air correction is the biggest one, and the most important, that's why it is needed to calculate it as good as possible.

Bouguer correction is the adjustment to a measurement of gravitational acceleration to account for elevation and the density of rock between the measurement station and a reference level. It compensates for the gravitational attraction of a plate of constant thickness h. This attraction is calculated by formula:

$$\delta g_B = 0,4186\rho h \tag{3}$$

where ρ is the assumed average plate density of subsoil and h is the plate thickness calculated above the reference ellipsoid. A value of thickness can be replaced and also used as relative value.

Terrain corrections are always taken to the distance of 166,735 km from the measured point. These corrections are replacing effect of valleys or mountains that was eliminated by previous Bouguer correction. In some cases, like this research, it is not necessary to calculate with the influence of valleys or mountains. In microgravity research is the most important correction the effect of sites' walls.

3   MEASURED AREAS

The measurement was done first on testing area. In that case, there was measured the influence of the cellar in urban setting, where we have known the shape of the searched cellar, also known the possible influence of this cellar on gravity. The details and results are possible to be seen in article [2]. After agreement with the historians, we decided to choose the first area, used for application of the method. The measurement was conducted in Saint Mary church in Kláštor pod Znievom village. The church was built as the part of monastery in 1251. The measurement was done with not regular density in whole church. The points of measurement are shown by crosses at Figure 3. The reason of this decision was because of the importance of the church parts, the more important areas were measured in higher density (presbytery, chapel), the less important areas were measured in lower density (nave, sacristy).

The result after all calculation showed us two places with negative Bouguer anomaly. The negative anomaly can be interpreted as the places, where subsoil has less density, or as there are free spaces in the subsoil. There is unwritten law, that the only measurement is no measurement. That was the reason, why was measurement also done with GPR system. The results from GPR survey confirmed, that there are two crypts and their sides were filled by sarcophagus. The conclusion was confirmed by endoscopic camera, there was made two holes. The first was made for the lightning and the second one for the camera.

This method was also used in historical site in Banská Belá village. There are two churches next beside. The measurement showed that the combination of these two methods is good enough, fast enough and precise to be used in urban settings.

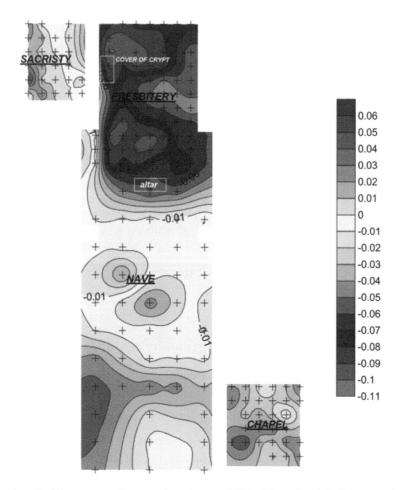

Figure 3.   The final Bouguer gravity map of presbytery of Saint Mary church in Kláštor pod Znievom (The darker places express the absence of anomaly, that could be interpreted as low subsoil density or as a free space in subsoil).

136

4 CONCLUSION

There is bigger importance of underground cavities searching, not only because of their archeologic importance, as is shown in paper, but because of geological questions as well. The microgravity method is possible to be used not only in opened area, but after right calculations and correction (terrain corrections), this method can be used in urban area as well. In good combination of geophysical methods, they are giving sureness of existence of the underground anomalies.

ACKNOWLEDGEMENTS

This article is the result of the implementation of the project VEGA 1/0275/17 "Application of numerical methods to define the changes of geometrical track position" supported by the Scientific Grant Agency of the Ministry of Education, science, research and sport of the Slovak Republic and the Slovak Academy of Sciences.

REFERENCES

[1] Butler, Dwain K., Microgravimetric and gravity gradient techniques for detection of subsurface cavities, GEOPHYSICS, 49(7), 1984.

[2] Chromčák J., Grinč M., Pánisová J., Vajda P., Kubová A., Validation of sensitivity and reliability of GPR and microgravity detection of underground cavities in complex urban settings: Test case for a cellar, 11th Slovak geophysical conference, Bratislava, pp 18–19, 2015.

[3] Mochales T., Casas A. M., Pueyo E. L., Pueyo O., Román M. T., Pocoví A., Soriano M. A., Ansón D., Detection of underground cavities by combining gravity, magnetic and ground penetrating radar surveys: a case study from the Zaragoza area, NE Spain, Environmental Geology, pp 1067–1077, January 2008.

[4] Tuckwell G., Grossey T., Owen S., Stearns P., The use of microgravity to detect small distributed voids and low-density ground, Quarterly Journal of Engineering Geology and Hydrogeology, The Geological Society of London, United Kingdom, pp 371–380 2008.

[5] Chromčák J., Pisca P., Grinč M., Microgravity and GPR Detection of underground Cavities in historical Sites, 16 International Multidisciplinary Scientific GeoConference SGEM, Albena, 2016.

[6] Lowrie W., Fundamentals of Geophysics, Cambridge University Press, United Kingdom, 2004, ISBN 0-521-46164-2.

[7] Pašteka, R., Zahorec P., Interpretation of microgravimetral anomalies in the region of the former church of St. Catarine, Dechtice, Contributions to Geophysics and Geodesy. Vol. 30. 4, 373–387.

[8] Pánisová, J., Pašteka, R., 2009: The use of microgravity technique in archaeology: A case study from the St. Nicolas Church in Pukanec, Slovakia. Contributions to Geophysics and Geodesy, Vol. 39, No. 3, 237–254.

Advances and Trends in Geodesy, Cartography and Geoinformatics – Molčíková et al. (Eds)
© *2018 Taylor & Francis Group, London, ISBN 978-1-138-58489-1*

Kronstadt height datum on the territories of the Czech and Slovak republics and its connection to a global vertical reference frame

V. Vatrt, R. Machotka & M. Buday
Faculty of Civil Engeneering, Brno University of Technology, Brno, Czech Republic

ABSTRACT: The development of Global Vertical Reference Frame (GVRF) based on a W_0 value is an actual topic to be solved by the IAG (International Association of Geodesy). In the past was developed methodology for determining the vertical shift of Local Vertical Datums over areas covered by GNSS/levelling sites. This methodology was applied to the Kronstadt Height Datum (KHD) on the territory of the Czech and Slovak Republics. Using the gravity field models EGM2008 and EIGEN-6C4 and the four primary constants (GM – geocentric gravitational constants, ω – nominal value of the angular velocity of the Earth's rotation, J_2 – the second zonal Stokes geopotential harmonic coefficient, W_0 – geoidal potential), the vertical shift of KHD to actual geopotential value W_0 = 62 636 856.0 m² s⁻² and 62 636 854.0 m² s⁻² has been determined—range from −2.2 cm to +1.5 cm, resp. −23.0 cm up to −19.3 cm for first, respectively second value W_0.

1 INTRODUCTION

In the past, the methodology for geopotential model testing was developed (Burša et al., 1997). As a by-product of geopotential models testing also the technology for determining the geopotential values at the tide gauge stations was developed. The tide gauge stations are used for specifying the Local Vertical Datums (LVDs) over areas covered by GNSS/levelling sites (Burša et al., 2001).

The mentioned methodology was many times applied (Burša et al., 1999a, 1999b, 2001, 2002a, 2002b). In this paper, the technology is applied repeatedly for territories of the Czech Republic and Slovak Republic, the area for testing geopotential models (TGM) has been the same but using significantly higher density of available GNSS/levelling sites.

1.1 Testing geopotential models

The methodology of testing geopotential models (TGM) is based on the Molodensky's theory (Molodensky et al. 1962) which says by Fig. 1 that there is a such point N on the normal plumb-line of any site M at the actual physical Earth's surface.

$$U(N) = W(M), \tag{1}$$

where the value of actual geopotential $W(M)$ generated by the Earth's body at GNSS/levelling point M on the Earth's surface, $U(N)$ is the normal potential generated by normal ellipsoid which is defined by four fundamental constants (8)–(12). The position of the point N on the normal plumb-line is determined by the Molodensky's normal height H_q which is available with an accuracy limited by the levelling errors only. The spatial resolution of Global Gravity Models (GGM), such as EGM2008 (Pavlis et al., 2012) and EIGEN-6C4 (Shako et al., 2013), does not provide the local anomalies of the gravity field. Due to this fact the equation (1) is transformed into,

$$U(N) \neq W(M), \tag{2}$$

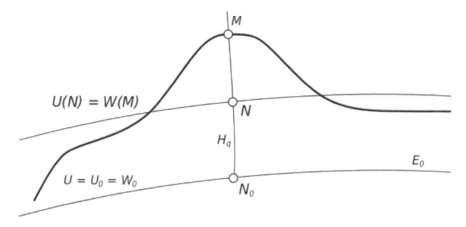

Figure 1.   Theory of geopotential model testing.

Then

$$\delta W = U(N) - W(M) \text{ and } \delta R = -\frac{GM}{W^2}\delta W \tag{3}$$

are the distortions δW or radial the distortions δR. For this paper the values of geopotential are obtained from EGM2008 and EIGEN-6C4. The values of these distortions δW or δR are small if the following quantities are accurate enough:

– fundamental constants GM, ω, W_0, J_2
– geocentric coordinates
– normal Molodensky's heights H_q of the testing sites.

The errors in geocentric coordinates $(\varphi, \lambda, h_{el})$ and H_q are at level of a few centimeters, the height differences can be caused by different Local Vertical Datums (LVD) that are used. That is why there are non-zero biases in δW or in δR (3).

Standard deviation of tested geopotential model obtained over an area covered by testing sites is

$$\sigma_{mod} = \sqrt{\frac{\sum\limits_{i'=1}^{n}(\overline{\delta R} - \delta R_i)^2}{n-1}}, \text{ where } \overline{\delta R} = \frac{\sum\limits_{i=1}^{n}\delta R}{n}. \tag{4}$$

2   BASIC PRINCIPLES FOR CONNECTING THE LVDS TO GLOBAL VERTICAL REFERENCE FRAME (GVRF)

GVRF should be determined on the basic geopotential value W_0 which has been developed on the basis of TOPEX/POSEIDON and JASON 1, 2 altimeter data (Burša et al., 1999b, 2002a, 2002a, 2002b, 2005, 2012). The mean rounded values of W_0 have been adopted as (Burša et al. 1998.):

$$W_0 = 62\ 636\ 856.0\ \text{m}^2\ \text{s}^{-2} \tag{5}$$

and its refined rounded value as (Sánchez et al. 2016):

$$W_0 = 62\ 636\ 854.0\ \text{m}^2\ \text{s}^{-2}. \tag{6}$$

After developing the W_0 values (5) and (6) the connection of the LVDs to GVRF can be realized through the differences

$$\Delta W_{0,i} = W_0 - W_{0,i} \tag{7}$$

between W_0 and the geopotential value $W_{0,i}$ at each i-th tide gauge station, defining the i-th LVD. Difference (7) can be determined on the basis of the GNSS/levelling data of the precise geopotential models (EGM2008, EIGEN-6C4), and the fundamental constants:

– geocentric gravitational constants (Ries et al., 1992).

$$GM = (398\ 600\ 441.8 \pm 0.8) \times 10^6\ m^3 s^{-2} \tag{8}$$

– nominal value of the angular velocity of the Earth's rotation

$$\omega = 7\ 292\ 115 \times 10^{-11}\ rad \cdot s^{-1} \tag{9}$$

– the second zonal Stokes geopotential harmonic coefficient (IAG SC3 Fin. Rep., 1995).

$$J_2 = (1\ 082\ 635.9 \pm 0.1) \times 10^{-9}; (\text{ zero-frequency tide system}) \tag{10}$$

$$J_2 = (1\ 082\ 666.7 \pm 0.1) \times 10^{-9}; (\text{ mean tide system}) \tag{11}$$

$$J_2 = (1\ 082\ 626.7 \pm 0.1) \times 10^{-9}; (\text{ tide-free}). \tag{12}$$

Note that the three values of J_2, Eq. (10), (11) and (12) should be used correctly by used tide reference systems of GNSS coordinate reference system, mean sea levels and geopotential models. The values (10)–(12) and the three adopted fundamental constants (5) or (6), (8), (9) define the mean Earth's level ellipsoids (E_0) parameters a, α (the semimajor axis and the flattening) by Pizzetti's theory (Pizzetti, 1913) and by modified Pizzett's theory (Burša et al., 1999b).

The three surfaces E_0, specified by the parameters Table 1 or Table 2, are required for unifying the Molodensky's normal heights with regards to the tide reference systems.

The parameters of E_0, Molodensky's normal heights and precise geopotential models (here EGM2008, EIGEN-6C4) make it possible to compute the actual geopotential $W(M)$ at any levelling point M on the Earth's physical surface and normal potential $U(N)$ at point N (Fig. 1) of E_0. For each point of testing area the difference can be computed from equation (3).

Final vertical shift $\Delta W_{0,i}$ (7) of LVD to GVRF in $m^2 \cdot s^{-2}$ and $\Delta H_{0,i}$ in meters is obtained from:

Table 1.   Parameters of the level ellipsoid E_0 in zero, mean, and tide-free systems for adopted fundamental constants and $W_0 = 62\ 636\ 856.0\ m^2 \cdot s^{-2}$ (Sánchez, 2016).

Tidal system	$J_2 [10^{-6}]$	a[m]	$1/\alpha$
Zero tide	1082.6359	6378136.58	298.25645
Mean tide	1082.6667	6378136.68	298.25233
Tide free	1082.6267	6378136.55	298.25768

Table 2.   Parameters of the level ellipsoid E_0 in zero, mean, and tide-free systems for adopted fundamental constants and $W_0 = 62\ 636\ 854.0\ m^2 \cdot s^{-2}$ (Sánchez, 2016).

Tidal system	$J_2 [10^{-6}]$	a[m]	$1/\alpha$
Zero tide	1082.6359	6378136.79	298.25645
Mean tide	1082.6667	6378136.88	298.25234
Tide free	1082.6267	6378136.76	298.25768

$$\Delta W_{0,i} = \frac{\sum_{j=1}^{n} \delta W_j}{n} \pm \sigma_{\delta W} \qquad (13)$$

$$\delta H_{0,i} = \frac{\sum_{j=1}^{n} \delta R_j}{n} \pm \sigma_{\delta R}, \qquad (14)$$

where n is number of GNSS/levelling points.

However, the $W(M)$ values are distorted because of EGM2008 or EIGEN-6C4 model errors. That is why averaging $\Delta W_{0,i}$ over a "sufficiently large" area is absolutely necessary.

3 PRACTICAL REALIZATION OF KHD CONNECTION TO GVRF ON THE TERRITORIES OF THE CZECH AND SLOVAK REPUBLICS

A GVRF, specified by a reference value (5) or (6), should serve as a basis for a connection of KHD on the territories of the Czech and Slovak Republics to GVRF. The GNSS/levelling data, with higher density and accuracy are currently available on both territories, and this makes it possible to implement the TGM method. Fig. 2 shows the distribution of points that were used for computation of the connection and also for statistical confirmation for both territories.

The numerical results are shown in Tables 3, 4, 5, 6 where EGM2008R and EIGEN-6C4R are estimated using the EGM2008 and EIGEN-6C4 resolutions over the area covered by the available GNSS/levelling sites, evaluated by analogy of (NRC, 1997), $W_{0,i}$ – the geopotential value at the Kronstadt LVD-origin, and $\delta H_{0,i}$, which responds for the vertical shift of the LVD-origin, related to our adopted reference equipotential surface (5) or (6) i.e. $W = W_0$. Four independent solutions for KHD were provided in three variants representing an opportunity to estimate the actual accuracy that can be achieved by the methodology applied here on the territories the Czech and Slovak Republic.

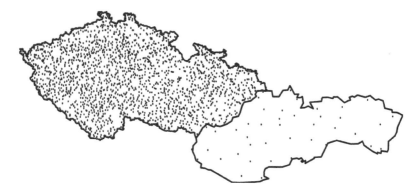

Figure 2.   Distribution of the GNSS/levelling points on the territories of the Czech and Slovak Republic.

Table 3.   Geopotential values W_{0i} at LVD's; δH_{0i} is the vertical shift of the LVD—origin, related to reference surface $W = W_0 = 62\,636\,856.0$ m^2 s^{-2}. TGM_ACC is accuracy of geopotential model; EGM2008R is the estimated resolution error of EGM2008.

Territory	Number of GNSS points	EGM2008R [cm]	TGM_ACC [cm]	W_{0i} [m^2.s^{-2}]	$W_{0i} - W_0$ [m^2.s^{-2}]	δH_{0i} [m]
Czech Rep.	2187	1.2	3.7	$62\,636\,856.22 \pm 0.12$	$+0.22 \pm 0.12$	-0.022 ± 0.012
Slovak Rep.	43	1.6	3.1	$62\,63\,856.07 \pm 0.17$	$+0.07 \pm 0.17$	-0.007 ± 0.017
combined	2230	1.0	3.7	$62\,636\,856.21 \pm 0.10$	$+0.21 \pm 0.10$	-0.022 ± 0.010

Table 4. Geopotential values W_{0i} at LVD's; δH_{0i} is the vertical shift of the LVD—origin, related to reference surface $W = W_0 = 62\,636\,854.0\ \text{m}^2\ \text{s}^{-2}$. TGM_ACC is accuracy of geopotential model; EGM2008R is the estimated resolution error of EGM2008.

Territory	Number of GNSS points	EGM2008R [cm]	TGM_ACC [cm]	W_{0i} [m².s⁻²]	$W_{0i} - W_0$ [m².s⁻²]	δH_{0i} [m]
Czech Rep.	2187	1.2	3.7	62 636 858.26 ± 0.12	+2.26 ± 0.12	−0.230 ± 0.012
Slovak Rep.	43	1.6	3.7	62 636 868.11 ± 0.18	+2.11 ± 0.17	−0.215 ± 0.017
combined	2230	1.0	3.7	62 636 858.26 ± 0.10	+2.26 ± 0.10	−0.229 ± 0.010

Table 5. Geopotential values W_{0i} at LVD's; δH_{0i} is the vertical shift of the LVD—origin, related to reference surface $W = W_0 = 62\,636\,856.0\ \text{m}^2\ \text{s}^{-2}$. TGM_ACC is accuracy of geopotential model; EIGEN-6C4R is the estimated resolution error of EIGEN-6C4.

Territory	Number of GNSS points	EIGEN-6C4R [cm]	TGM_ACC [cm]	W_{0i} [m².s⁻²]	$W_{0i} - W_0$ [m².s⁻²]	δH_{0i} [m]
Czech Rep.	2187	1.2	3.7	62 636 856.18 ± 0.12	+0.18 ± 0.12	−0.019 ± 0.012
Slovak Rep.	43	1.6	3.1	62 63 855.86 ± 0.16	−0.15 ± 0.17	0.015 ± 0.017
combined	2230	1.0	3.7	62 636 856.18 ± 010	+0.18 ± 0.10	−0.018 ± 0.010

Table 6. Geopotential values W_{0i} at LVD's; δH_{0i} is the vertical shift of the LVD—origin, related to reference surface $W = W_0 = 62\,636\,854.0\ \text{m}^2\ \text{s}^{-2}$. TGM_ACC is accuracy of geopotential model; EIGEN-6C4R is the estimated resolution error of EIGEN-6C4.

Territory	Number of GNSS points	EIGEN-6C4R [cm]	TGM_ACC [cm]	$W_{0,i}$ [m².s⁻²]	$W_{0i} - W_0$ [m².s⁻²]	δH_{0i} [m]
Czech Rep.	2187	1.2	3.7	62 636 858.23 ± 0.12	+2.23 ± 0.12	−0.226 ± 0.012
Slovak Rep.	43	1.6	3.7	62 636 857.90 ± 0.16	+1.90 ± 0.17	−0.193 ± 0.017
combined	2230	1.0	3.7	62 636 858.22 ± 0.10	+2.20 ± 0.10	−0.226 ± 0.010

4 CONCLUSIONS

A global vertical reference system can be defined by a choice of a W_0 value (5), (6), the adopted reference value of W_0 can, in general, be chosen arbitrarily.

The vertical shifts between the LVDs-origin KHD and GVRS on the territory of the Czech Republic and Slovak Republic are:

- for EGM2008 and W_0 (5) is about −2.2 cm, vertical shift of combined solution (both territories) is close to solution for the Czech Republic only (Table 3)
- for EIGEN-6C4 and W_0 (5) is only about −1.8 cm, vertical shift of combined solution (both territories) is of close to solution of the Czech Republic (Table 3)
- for EGM20808 and W_0 (5) and EIGEN-6C4 and W_0 (5) is from 0.3 cm to 0.8 cm for all variant of territories (Table 3 and Table 4)
- for EGM2008 and W_0 (6) is about −23.0 cm, vertical shift of combined solution (both territories) is close to solution for the Czech Republic only (Table 5)
- for EIGEN-6C4 and W_0 (6) is only about −22.6 cm, vertical shift of combined solution (both territories) is close to solution for the Czech Republic only (Table 6)
- for EGM20808 and W_0 (6) and EIGEN-6C4 and W_0 (6) is from −23.0 cm to −20.0 cm for all variant of territories (Table 5 and Table 6).

The combined solution for both territories is more precise then separate one. The vertical shift of the LVD-origin KHD to GVRF is from −2.2 cm to +1.5 cm for W_0 (5) and from −23.0 cm up to −19.3 cm for W_0 (6). The final solution is heavily influenced by different density of GNSS/levelling points that were available for this study for each country. The final solution should have the same distribution of stations for both countries and the most preferable way is to cover the other countries that also use the LVD KHD.

ACKNOWLEDGEMENT

The authors would like to thank the Research Institute of Geodesy and Cartography in Bratislava for providing the GNSS/levelling stations data from territory of the Slovak republic.

REFERENCES

Burša, M., Raděj, K., Šíma, Z, True, S.A., Vatrt, V. (1997). Test for accuracy of recent geopotential models. *International Geoid Service Bulletin No 6,* D.I.I.A.R. Politecnico di Milano, Italy, 1997, p. 167–188.

Burša, M., Groten, E., Kenyon, S., Kouba, J., Raděj, K., Vatrt, V., Vojtíšková, M (2002b). Earth's dimension specified by geoidal geopotential. *Studia geophysica et geodaetica*, **46**, 2002, p. 1–8.

Burša, M., Kenyon, S., Kouba, J., Raděj, K., Šimek, J., Vatrt, V., Vojtíšková, M. (2002a). World Height System Specified by Geopotential at Tide Gauge Stations. *IAG Symp. on Vertical Reference Systems*, Cartagena, Colombia, Feb. 20–23, 2001. Springer Verlag, 2002.

Burša, M., Kenyon, S., Kouba, J., Šíma, Z., Vatrt, V., Vítek, V., Vojtíšková, M. (2007). The Geopotential Value W_0 for Specifying the Relativisti Atomic Scale and Global Vertical Reference Frame. *Journal of Geodety*, **81**, 2007, p. 103–110.

Burša, M., Kouba, J., Kumar, M., Müller, A., Raděj, K., True, S.A., Vatrt, V., Vojtíšková, M. (1999b). Geoidal Geopotential and World Height System. *Studia geophysica et geodaetica*, 43, 1999, p. 327–337.

Burša, M., Kouba, J., Raděj, K., True, S.A., Vatrt, V., Vojtíšková, M. (1998). Mean Earth's equipotential surface from TOPEX-POSEIDON altimetry. *Studia geophysica et geodaetica*, 42, 1998, p. 459–466.

Burša, M., Kouba, J., Raděj, K., True, S.A., Vatrt, V., Vojtíšková, M. (1999a). Determination of the geopotential at the tide gauge defining the North American Vertical Datum 1988 (NAVD 88). *Geomatica*, 53,, no. 3, 1999, p. 157–162.

Burša, M., Kouba, J., Müller, A., Raděj, K., True, S. A., Vatrt, V., Vojtíšková, M. (2001). Determination of geopotential differences between local vertical datums and realization of a World Height System. *Studia geophysica et geodaetica*, 45, 2001, p. 127–132.

IAG SC3 Final Report, Travaux de L'Association Internationale de Géodésie,1995. 30, 370–384, IAG, Paris.

Molodenskij, M.S., Jeremejev, B.F., Jurkina, M.I. (1962). Methods for study of the external gravitational field and figure of the Earth. Israel program for scientific translations, Jerusalem (translated from Russian original, Moscow 1960).

NRC, SATELLITE GRAVITY AND GEOSPHERE (1997). Contributions to the Study of the Solid Earth and Its Fluid Envelope, Commision on Geosciences, Enviroment, and Recources, Chair Jean Dickey, National Academy Press, Washington, D.C., pp. 112, 1997.

Pavlis, N.K., Holmes, S.A., Kenyon, S.C., Factor, J.K. (2012). The development and evaluation of the Earth Gravitational Model 2008 (EGM2008), Journal of Geophysical Research: Solid Earth (1978–2012) Volume 117, Issue B4, April 2012.

Pizzetti P., 1913: Principii della teoria meccanica della figura dei pianeti. Pisa. E. Spoerri, XIII, 251 pp.

Ries, J.C., Eanes, R.J., Shum, C.K., Watkins, M.M. (1992). Progress in the determination of the gravitational coefficient of the Earth. Geophys. Res. Letters, 19, No. 6, 271–274, 1992.

Sánchez, L., Čunderlík, R., Dayoub, N., Mikula, K., Minarechová, Z., Šíma, Z., Vatrt, V., Vojtíšková, M. (2016). A conventional value for the geoid reference potential W0. *Journal of Geodesy*, *90*(9), 815–835.

Shako, R., Förste, C., Abrikosov, O., Bruinsma, S., Marty, J.-C., Lemoine, J.-M., Dahle, C. (2013). EIGEN-6C: A High-Resolution Global Gravity Combination Model Including GOCE Data. Observation of the System Earth from Space – CHAMP, GRACE, GOCE and Future Missions, 155–161. doi:10.1007/978-3-642-32135-1_20.

Part C: Cartography and geoinformatics

Advances and Trends in Geodesy, Cartography and Geoinformatics – Molčíková et al. (Eds)
© 2018 Taylor & Francis Group, London, ISBN 978-1-138-58489-1

Some possibilities of using geographic information systems in analysis of the potential of destination Slovenský raj (Slovakia) in tourism

P. Blišťan & M. Šoltésová
Faculty of Mining, Ecology, Process Control and Geotechnology, Institute of Geodesy, Cartography and Geographical Information Systems, Košice, Slovakia

B. Kršák, C. Sidor & Ľ. Štrba
Faculty of Mining, Ecology, Process Control and Geotechnology, Institute of Earth Resources, Košice, Slovakia

ABSTRACT: Tourism in Slovakia represents the potential that in the future can positively influence the economy of Slovakia by bringing tourists and new services to the regions and creating new job opportunities. Several steps must be taken to achieve these goals, one of which is to analyze the potential of tourism objects. Geographic Information Systems (GIS) are often used in the process of identifying and analyzing objects of tourism. This paper presents some possibilities of using GIS in identifying and analyzing the potential of the Slovak Paradise (Slovakia) site for tourism. The results of the analyses are then usable, e.g. for marketing or strategic decision making in investing in the construction of new tourism objects.

1 INTRODUCTION

In a modern society, information technology plays a key role in the decision-making process. The most advanced countries in the world have reached their advanced economic level thanks to investments in information systems. Geographic Information Systems (GIS) is a specific category of information systems. They are used to collect, process, analyze, and present spatially localized data. Their application is broad, from the environment, through trade and logistics, to state and public administration. Their use is also in the field of resource management and analysis of the efficiency of their use, also in connection with tourism. This article concerns the area of tourism. It presents an example of GIS usage in the field of potential analysis and planning of capacity utilization of the Slovenský raj in tourism.

2 MATERIAL AND METHODS

Spatial analyses are one of the GIS tools that is used to assess the potential of a territory. We can identify areas of concentration considering occurrence of selected phenomena and evaluate their spatial connection within the frame this group of analyses.

2.1 *Spatial analysis in GIS*

Spatial analyses can be defined as a set of techniques for analysis and modeling of localized objects and phenomena. The results of the analyses depend on the spatial arrangement of these objects or phenomena and their properties (Horak, 2015). Spatial analyses require access to the attributes of studied objects, including location information. Spatial data analyses are connected with studies of spatial data system, with the aim to understand their dividing regularities. Based on this knowledge, they allow to predict development of studied phenomenon in the interest area (Horak, 2015). Spatial analyses includes statistical characteristics derivation of geo-elements' observed texture (points, lines or grounds), testing randomness distribution of their spatial allocation, seek for links between entities and description of the development studied phenomenon in unknown locations—interpolation.

2.1.1 *Hot spot analyses*

Hot spot (places with increased intensity of the examined phenomenon) are one of analysis forms of phenomena concentration in space (Fig. 1). In terms of destination visit rate analysis, these are areas with more than above-average number of visitors in the examined destination's area. In these destinations, it is necessary to focus on quality and complexity of services, ensure the absent services and actively work with client's satisfaction. Hot spot analysis' methods identify areas with high potential and visitors' interest in tourism objects. These objects are not evenly distributed in space. Their occurrence is conditioned by natural conditions and the presence of tourist objects such as hotels or events to attract visitors. It is possible to track several "parameters" of visitors—age, gender, residence, etc. A sufficient amount of data is needed to investigate this visitor information (Perry, 2013).

Execution of hot spot analyses is the first step in the geomarketing strategy. High potential sites will be identified with priority being given to financial resources to support tourism. There are several hotspot analysis techniques. Thematic mapping of administrative territories, quadrant method and nuclear estimates using the Kernel density function are the most commonly used methods (Chainey et al., 2008).

Quadrant method

Basis of the quadrant method is monitoring of events frequency in defined cells. Different methods with regular and irregular grid are used for transformation. The number of events belonging to individual cells indicates the value of continuous surface at given site (Horák, 2011). Cells showing high values of the observed phenomenon and identified clusters can be referred-to as a hot spot. It is important to determine the appropriate cell size that predestines spatial resolution analysis to obtain the correct results of the quadrant method.

Kernel density estimation

The Kernel Density Estimation (KDE) is a method of kernel estimation that uses point data in given area to analyze intensity of observed phenomenon visualized by a continuous field (Fig. 2). The analysis evaluates distance and statistical significance of individual points in relation to surrounding points, depending on the justified distance of the band. A very bland grid is created above the point field. At each of the points in this grid, the contribution of individual events is calculated using the function of the nuclear estimation. For each cell in the grid (raster), the sum of overlays is calculated to quantify the significance of phenomenon at specific locations in investigated space. It is also possible to use nuclear estimates that calculate with weights, which affects the analyzed phenomenon. For example, population density, etc. (Perry, 2013).

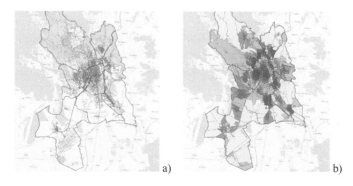

a)   b)

Figure 1.   Hot spot analysis principle – a) input data, b) output.

2.2 Case study—identification and analysis of the Slovenský Raj's potential

The Slovenský Raj National park is situated in the eastern part of Slovakia. It is located in the Spiš region and borders the Gemer region (Fig. 3). Its territory is situated on cantons Spišská Nová Ves, Rožňava, Poprad and Brezno. Together with the protection zone, 12 municipalities are involved. The natural area's conditions give opportunities for creating a complex year-round tourism offer with utilizing the potential of various natural and cultural-historical components. Based on the unique potential of territory, it is possible to adapt the leisure activities and the offered tourism products to different age categories and groups of visitors (e-volution, 2014).

In this area of interest, selected GIS tools used to analyze the tourist potential of the destination will be demonstrated.

2.3 Identification of tourism potential in the area of interest

The development of tourism is conditioned by a number of factors entering to create an offer:

- relief,
- waters,
- flora and fauna,
- climate,
- cultural and historical heritage.

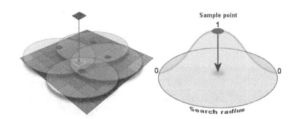

Figure 2. Kernel method principle.

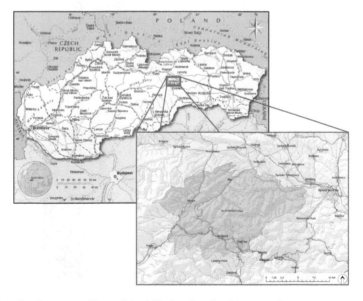

Figure 3. Localization map—Slovenský raj National park with protective zone.

These factors are given and unchangeable and constitute a basic prerequisite for the area's interest from the point of view of tourism. Objects of the secondary potential of territory mostly include:

- active sports and tourism,
- accommodation,
- catering.

2.3.1 *The primary potential of tourism*
Natural conditions
The national park is characterized by typical karst relief. The deep gorges, water streams, waterfalls and caves make the Slovenský Raj special within the frame of all national parks of Slovakia. Natural phenomenas such as Dobšinská ľadová jaskyňa (Fig. 4), Čertova Diera, Stratenská jaskyňa, Medvedia jaskyňa and Hranovnické Pleso are the famous attractions of the National park. Tomášovský výhľad (Fig. 5), Dobšinská ľadová jaskyňa, Prielom Hornádu and Suchá Belá gorge belongs to the most visited locations. There are 11 national nature reserves and 9 nature reserves within territory of the National park.

Hiking trails and biking trails
High concentration of technical security equipment on individual tourist routes is a distinctive feature of the Slovenský raj National park. On average, 1 km of hiking trails falls on 1 km^2 Slovenský Raj National park's area. Total monitored length of tourist routes is 458,8 km. Cycling can be done on local highways and tertiary roads. There are almost 325,2 km of cycling routes in the Slovenský raj destination and in its vicinity. For the cyclists, there are attractive locations such as Čingov, Podlesok, Píla, Kláštorisko, Dobšinská Ľadová Jaskyňa – Stratená, Dedinky, Mlynky, Hnilec, Spišská Nová Ves (e-volution, 2014).

Figure 4.   Dobšinská ľadová jaskyňa.

Figure 5.   Tomášovský výhľad.

Educational trails

Educational trails allow to discover the beauty and natural values of the area without local tour guide. There were built up to 12 educational trails in Slovenský raj National park and its protection zone till now. The total length of educational trails is almost 87 km.

Tourist resorts

Tourist resorts in national park can be allocated from the perspective of areas with a higher occurrence of tourist services and attractions. Significant places from this point of view are Podlesok, Čingov, Kláštorisko, Dobšinská ľadová jaskyňa, Dedinky and Mlynky. Košiarny briežok, Novoveská Huta, Píla, Geravy are also available in combination with others transport options.

2.3.2 *Secondary tourism potential*

Accommodation

Information about accommodation capacities in Slovenský raj destination were obtained through a survey of publicly available data in 2014. 232 operated accommodation establishments of different categories were identified with capacity more than 4.600 beds. These are facilities that are officially used for business activities (e-volution, 2014).

Catering

There is 84 catering establishments of various categories located in territory of Slovenský raj and in the vicinity in current period. More than 15 establishments of this number belong to the accommodation facilities. A total capacity has been identified for more than 6.000 places. Because it was not possible to identify the number of places for all catering establishments, the total capacity may vary. In addition, considering relatively easy entry into the business sector, these numbers can be significantly changed in a relatively short period of time by the influx of new entrepreneurs or the disappearance of existing ones (e-volution, 2014).

2.4 *Analysis of transport accessibility of the Slovenský raj's area*

The location of destination can be called centralized in regard of entire territory of the Slovak Republic. Slovenský raj is available via road networks, including a partial highway connection and railway tracks (Fig. 6). From the point of view of local accessibility, there are regular public transport service lines—regular intercity bus and rail transport. From the air transport point of view, the international airports Poprad – Tatry and Košice are situated nearest.

2.5 *Spatial analysis of tourism objects in GIS*

Spatial analyses of tourism's objects belongs, considering the demand for data quality, to the most demanding analyses. The spatial analyses, most commonly used for tourism purposes,

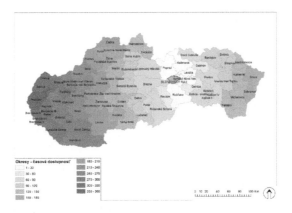

Figure 6.   GIS analysis of destinations time availability from individual districts of Slovakia.

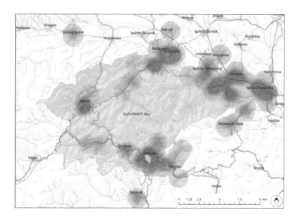

Figure 7. Concentration of accommodation capacities at destination made in GIS Kernel analysis.

include spatial statistics, ambient and network analysis, but also interpolation methods. In the following example we want to present the use of interpolation methods in the process of assessing the concentration of tourism objects. The spatial distribution of accommodation facilities was also analyzed in terms of spatial distribution—concentration of facilities in the area of interest. Nuclear estimation with Kernel method was also used to analyze data and identify locations with a high concentration of accommodation facilities. Based on the result of the analysis (Fig. 7), a higher concentration of accommodation facilities in the northern part of national park can be stated. However, it should be noted that the categories of accommodation capacity are distributed relatively equally. High proportion of private chalets is characteristic for the monitored area. Each of the entrance gates is also a place with a higher concentration of different accommodation capacities.

2.6 *Discussion and conclusions*

Tourism is a complex activity and requires the use of tools that will contribute to effective decision-making in assessing investment in this specific business area. GIS is a powerful and effective tool that can be well used to assess the area potential from the point of view of tourism. GIS can play an important role in the field of environmental auditing, in assessing the suitability of tourism sites but also in modeling spatial relationships. GIS is currently underused in Slovak tourism, with the exception of a few examples. The increasing interest in comprehensive information available on the web portals points to the need to build comprehensive databases with information about tourist objects in the Slovak Republic. These data can be used not only to improve tourists' awareness but also to various analyses in GIS for the purposes of assessing the effectiveness of investment in tourism.

REFERENCES

Chainey, S., Tompson, L. & Uhlig S. 2008. The utility of hotspot mapping for predicting spatial patterns of crime. Security Journal Vol. 21. pp 4–28.
Horák, J. 2011. Prostorové analýzy dat. Ostrava: Institut geoinformatiky, VŠB-TU Ostrava, 127 p.
Horák, J. 2015. Prostorové analýzy dat. Ostrava: Institut geoinformatiky, VŠB-TU Ostrava, 6. ed.
Perry, W.L. et al. 2013. Predictive Policing: The Role of Crime Forecasting in Law Enforcement Operations.
e-volution, 2014: Marketingová stratégia destinácie Slovenský raj. Košice, 218 p. [online]. [cit. 2011–20–07]. <http://www.vraji.sk/public/upload/files/Marketingova_strategia_destinacie_Slovensky_raj_2014.pdf>.
www.1: http://www.geography.hunter.cuny.edu/~jochen/GTECH361/lectures/lecture11/concepts/Kernel%20density%20calculations.htm.

Advances and Trends in Geodesy, Cartography and Geoinformatics – Molčíková et al. (Eds)
© *2018 Taylor & Francis Group, London, ISBN 978-1-138-58489-1*

Aggregation of uncertain information and its implementation in geographic information systems and spatial databases

R. Ďuračiová & M. Muňko
Department of Theoretecal Geodesy, Faculty of Civil Engineering, Slovak University of Technology, Bratislava, Slovak Republic

J. Caha
Department of Regional Development and Public Administration, Faculty of Regional Development and International Studies, Mendel University in Brno, Brno, Czech Republic

ABSTRACT: Geographic Information System (GIS) is a very useful tool for decision making using spatial data. Spatial decisions are usually made by aggregation of spatial or thematic criteria based on spatial data stored in databases. Spatial data are inherently uncertain, so it is also necessary to aggregate them by appropriately selected functions used for uncertain data. Uncertainty of spatial or thematic criteria can be efficiently expressed by fuzzy sets, and for aggregation of uncertain criteria, fuzzy logic operators (mainly the fuzzy AND, OR, and NOT operators, usually defined as the minimum, maximum, and complement) are commonly used. Other types of aggregation functions are means and averages, which can be also used for aggregation of criteria modelled by fuzzy sets. There is a lot of aggregation operators and each operator has its specific characteristic, so it is important to know which one to use for what purpose. We present the most used aggregation operators such as fuzzy logic operators represented by t-norms and t-conorms, quasi-arithmetic means (or generalized means), a weighted arithmetic mean, and an Ordered Weighted Averaging (OWA) aggregation operator. The result is a description of their implementation in GIS software environments and spatial database systems. The selected operators are applied in solving the tasks of spatial multi-criteria decision making and spatial predictive modelling. The advantages and disadvantages of their particular use are justified in individual cases.

1 INTRODUCTION

Although spatial data are often uncertain or vague, uncertain spatial queries and analysis of uncertain spatial data are still not sufficiently implemented in database systems and GIS. The functions and operations in database systems are commonly implemented by the use of crisp rules and criteria based only on Boolean logic. This classical approach can lead to a loss of information resulting from the uncertainty of the input data or decision criteria. For example, in the human perception is common to express quantitative characteristics using terms such as near, far, large, high, or low. The human mind is able to translate vague conditions and process them very well, but the computer can not process such queries. Therefore, it is first necessary to define uncertain terms and related functions and operators to make the system able to process them. In such cases, we can effectively apply fuzzy set theory. The concept of fuzzy sets and fuzzy logic was introduced by Professor Zadeh (1965). Fuzzy approach is quite commonly used in GIS in raster data analysis (Petry et al., 2005; Kacprzcyk et al., 2010; Shi, 2010; Shi et al., 2016), but not all useful aggregation operators are implemented in them. An example of the use of aggregation operations with fuzzy sets in GIS may be aggregation of spatial information generated as a result of analyses or expert estimates. In vector data analysis, mostly only bivalent Boolean logic is used (both in spatial database systems and in GIS). We can say that the basic principles of the fuzzy set theory, and in particular the aggregation operators (Gupta and Qi, 1991), (Grabisch et al., 2009), are not sufficiently implemented in current information systems. Therefore, this paper deals with the basic possibilities of implementation of the fundamental principles of fuzzy set theory (specifically aggregation operators) into uncertain spatial data queries and analysis. In next sections, we provide a brief overview of aggregation operators, which can be useful in uncertain spatial data analysis. We also briefly present currently implemented and commonly used aggregation operators in spatial data analysis. As a result, we propose some new solutions for implementation of various types of aggregation operators into GIS or spatial database systems.

2 METHODS

The fuzzy set theory is described in many books and papers and various aggregation operators are an important part of it. Therefore, as the basis for further implementation, the basic principles of fuzzy set theory, the short theory of the T-operators (T-norms and T-conorms), as well as definitions of means and averages are presented.

2.1 The basic principles of fuzzy set theory

Fuzzy set theory can be understood as an extension of the conventional set theory. The characteristic function $\chi_A(x)$ of a crisp set A, which corresponds to the use of Boolean logic, assigns a value of either 1 or 0 (Fig. 1). This function can be generalized in such a way that the generalised function $\mu_{A'}(x)$ takes values from the interval $\langle 0,1 \rangle$. Such a function is called a fuzzy membership function, and the set defined by it is called a fuzzy set A' (Fig. 1). Larger value denotes higher degree of membership to fuzzy set, which can represent, for example, a spatial criterion. Therefore, using the fuzzy membership function, we can express the fulfilment or non-fulfilment of the criterion in spatial analyses.

2.2 The AND, OR, and NOT operators in fuzzy logic (Some typical t-operators)

Fuzzy logic can be also considered as a generalization of Boolean logic. Unlike conventional logic, the fuzzy logical operators are defined in more than one way. The standard fuzzy logical operators were defined in (Zadeh, 1965). Zadeh's conventional operators, MIN and MAX, are used in almost every decision-making processes modelled by fuzzy sets. However, some studies indicate that other types of fuzzy logical operators may work better in some situations (Gupta and Qi, 1991). In general, the operations of fuzzy conjunction (AND) are expressed as triangular norms (t-norms) and operations of fuzzy disjunction (OR) are represented by triangular conorms (t-conorms) (Grabish et al., 2009).

   According to Gupta and Qi (1991), the t-norm and the t-conorm originated from the studies of probabilistic metric spaces (Menger, 1942; Schweizer and Sklar, 1983) in which triangular inequalities were extended using the theory of t-norms and t-conorms.

   The most widely used t-norms, t-conorms and complement $\bar{A}$ of the set A are defined as follows:

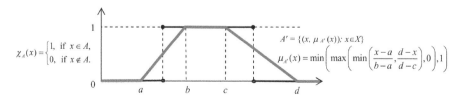

$$\chi_A(x) = \begin{cases} 1, & \text{if } x \in A, \\ 0, & \text{if } x \notin A. \end{cases} \qquad A' = \{(x, \mu_{A'}(x)); x \in X\} \qquad \mu_{A'}(x) = \min\left(\max\left(\min\left(\frac{x-a}{b-a}, \frac{d-x}{d-c}\right), 0\right), 1\right)$$

Figure 1.   Crisp set (A) defined by the characteristic function $\chi_A(x)$ versus trapezoidal fuzzy set (A') defined by the fuzzy membership function $\mu_A(x)$.

Zadeh's T-operators (Zadeh, 1965):

$$T_M(x,y) = \min(x,y), \quad \text{– the minimum t-norm (standard conjunction)} \tag{1}$$
$$S_M(x,y) = \max(x,y) \quad \text{– the standard t-conorm (maximum)} \tag{2}$$
$$\mu_{\bar{A}}(x) = 1 - \mu_A(x) \tag{3}$$

Probabilistic operators (Bandler and Kohout, 1980):

$$T_P(x,y) = xy, \quad \text{– the product t-norm} \tag{4}$$
$$S_P(x,y) = x + y - xy \quad \text{– the product t-conorm (probabilistic sum)} \tag{5}$$
$$\mu_{\bar{A}}(x) = 1 - \mu_A(x) \tag{6}$$

Łukasiewicz operators (Giles, 1976):

$$T_L(x, y) = \max(0, x + y - 1) \quad \text{– the Łukasiewicz t-norm} \tag{7}$$
$$S_L(x, y) = \min(1, x + y) \quad \text{–the Łukasiewicz t-conorm (bounded sum)} \tag{8}$$
$$\mu_{\bar{A}}(x) = 1 - \mu_A(x) \tag{9}$$

Drastic operators (Weber, 1983):

$$T_D(x,y) = \begin{cases} \min(x,y), & \text{if } \max(x,y) = 1 \\ 0, & \text{else.} \end{cases} \quad \text{– the drastic t-norm} \tag{10}$$

$$S_D(x,y) = \begin{cases} \max(x,y), & \text{if } \min(x,y) = 0 \\ 1, & \text{else} \end{cases} \quad \text{– the drastic t-conorm} \tag{11}$$

$$\mu_{\bar{A}}(x) = 1 - \mu_A(x) \tag{12}$$

Another set of t-operators can be defined, for example, as follows (Gupta and Qi, 1991):

$$T_X(x,y) = \frac{xy}{x + y - xy} \tag{13}$$

$$S_X(x,y) = \frac{x + y - 2xy}{1 - xy} \tag{14}$$

$$\mu_{\bar{A}}(x) = 1 - \mu_A(x) \tag{15}$$

More details on fuzzy logical operators are described in (Gupta and Qi, 1991) or (Grabish et al., 2009).

2.3 *Means and averages*

In addition to fuzzy logical operators, also other aggregation operators can be used in some cases of spatial analyses. The well-known subclasses of aggregation operators are means and averages. The most used of them can be effectively defined by the quasi-arithmetic mean (also known as the generalized mean) (Klir and Yuan, 1995):

$$h_\lambda(\alpha_1,...,\alpha_n) = \left(\frac{1}{n} \sum_{i=1}^{n} \alpha_i^\lambda \right)^{\frac{1}{\lambda}}, \tag{16}$$

where $\alpha \in \langle 0,1 \rangle, n \geq 2, \lambda \in R, \lambda \neq 0$.
The commonly used special cases of generalized means are:

- the arithmetic mean (for $\lambda = 1$),
- the quadratic mean (for $\lambda = 2$),
- the harmonic mean (for $\lambda = -1$),
- the geometric mean (for $\lambda \to 0$),
- the maximum (for $\lambda \to +\infty$),
- the mimimum (for $\lambda \to -\infty$).

Note, that he last two cases correspond to the Zadeh's conventional fuzzy logical operators.

The weighting of individual criteria is often used in the aggregation process. Therefore, the weighted arithmetic mean (average) h_w is defined as follows:

$$h_w(\alpha_1,...\alpha_n) = \sum_{i=1}^{n} w_i \alpha_i, \qquad (17)$$

where $n \in \mathbb{N}$ and $w = (w_1,...w_n) \in \langle 0,1 \rangle^n$ is a vector of weights fulfilling the condition: $\sum_{i=1}^{n} w_n = 1$.

Ordered weighted averaging operator (OWA-operator) $\overline{h_w}$ is also determined by a vector of weights $w = (w_1,...w_n) \in \langle 0,1 \rangle^n$ fulfilling the condition $\sum_{i=1}^{n} w_n = 1$:

$$h_w(\alpha_1,...\alpha_n) = \sum_{i=1}^{n} w_i \alpha_{p(i)}, \qquad (18)$$

where p is such a permutation of the indexes that $\alpha_{p(1)} \leq \alpha_{p(2)} \leq ... \leq \alpha_{p(n)}$.

3 IMPLEMENTATION

3.1 *Aggregation of uncertain information represented by raster data*

For raster data, some principles of fuzzy set theory as well as basic aggregation operators are implemented in some GIS. Such GIS environments are GRASS GIS, ArcGIS, IDRISI, etc. For example, tools such as Fuzzy Membership and Fuzzy Overlay are implemented in ArcGIS. However, no GIS software environment supports all above mentioned fuzzy aggregation operators. If we need to apply an aggregation operator that is not implemented in ArcGIS, we can use tools such as Cell Statistics or Raster Calculator (tools for map algebra realisation) to define it.

As a case study, Figure 2 shows the samples of the layers for archaeological predictive modelling (Lieskovský et al., 2015), which express meeting criteria c1 (suitable soil), c2 (suitable slope), and c3 (suitable distance to watercourses) resulting from the categorical data (left), coverages (middle), and distance from spatial objects (right), respectively.

Then, for example, the Łukasiewicz t-norm and t-conorm, which are not implemented in ArcGIS, can be written into Raster Calculator using the Con operator and formulas (7) and (8) (Fig. 3). In a similar manner, all of the above-described aggregation operators may be applied to aggregate criteria represented by raster data in GIS. In addition, to implement new tools based on these operators, the ArcGIS ModelBuilder software environment can be used effectively.

Figure 2. Raster data representing meeting uncertain criteria: thematic (left), derived from coverage (middle), based on distance from polyline object (right).

```
Con(("c_1"+"c_2"+"c_3"-2)>0, "c_1"+"c_2"+"c_3"-2, 0)
```

```
Con(("c_1"+"c_2"+"c_3"<1, "c_1"+"c_2"+"c_3", 1)
```

Figure 3.   Aggregation of criteria represented by raster data in Raster Calculator.

```
CREATE OR REPLACE FUNCTION public.generalized_mean(
    real[],
    real)
    RETURNS real
    LANGUAGE 'plpgsql'
    COST 100.0

AS $function$                           SELECT id,
DECLARE                                     fuzzy_AND_Min(m1,m2) as standard,
    s real := 0;                            fuzzy_AND_Drs(m1,m2) as drastic,
    x real;                                 generalized_mean(ARRAY[m1,m2],1) as arithmetic,
    n int := 0;                             generalized_mean(ARRAY[m1,m2],2) as quadratic,
BEGIN                                       generalized_mean(ARRAY[m1,m2],-1) as harmonic
    FOREACH x IN ARRAY $1               FROM fuzzy_tab;
    LOOP
        s := s + x ^ $2;
        n := n + 1;
    END LOOP;
    s = (s / n) ^ (1 / $2);
    RETURN s;
END;
$function$;
```

Figure 4.   Implementation of generalized mean as an aggregation operator in PostgreSQL database system (left) and SQL query using fuzzy logical operators (standard and drastic t-norm) and three types of means (arithmetic, quadratic, harmonic) (right).

3.2   *Aggregation of uncertain information represented by vector data and their attributes*

For querying uncertain vector data, spatial database systems can provide an appropriate solution. Although aggregation operators are not implemented in them, they can be created as an extension or new functions. As an example, we present application of the fundamental principles of fuzzy aggregation operators to querying spatial data in object-relational database system PostgreSQL with the PostGIS extension. We propose to create all needed aggregation operators as functions stored in database management system. The function by which we implemented the generalized mean is shown as an example in Figure 4 (left). The input of the presented function is the attribute value array and the λ coefficient, which determines which type of means we apply. Implementations of all above defined fuzzy logical operators can be realised in similar way and are described in papers (Ďuračiová, 2014) and (Ďuračiová and Faixová Chalachanová, 2018). The query using the generalized_mean function and newly created fuzzy logical operators (e.g. fuzzy_AND_min and fuzzy_AND_drs) is shown in Figure 4 (right).

4   CONCLUSIONS AND DISCUSSION

In this paper, we emphasize the fact that the way of aggregation of input layers in GIS is not uniquely given but provides several options. Aggregation operator affects the result, so its selection should be substantiated or tested by validation of the resulted models (in case of predictive modelling). The choice of the aggregation operator should mainly result from the type of the decision criteria (input layers) and the purpose, for which they are used. The basic methods of aggregation of input layers in predictive modelling by means of logical operators and averages are described, for example, in (Lieskovský et al., 2013). Criteria for selecting aggregation operator were summarized by Zimmermann (2001) as following: axiomatic strength, empirical fit, adaptability, numerical efficiency, compensation, range of compensation, aggregation behaviour, and required scale level of membership functions. In general, logical operators eliminate the impact of unfulfilled criteria. The averages allow some criteria to be missed, and these criteria only reduce the resulting value. Depending on the type of input layers, individual criteria may be aggregated either together (by one aggregation operator) or sequentially, with sequential application of various aggregation operators (i.e., not only one aggregation operator for all input layers). Implementation of new fuzzy functions and aggregation operators into spatial database systems and GIS, which is described in this paper, can help improve the process of spatial multiple criteria decision making based on uncertain data.

ACKNOWLEDGEMENT

This work was supported by Grant No. 1/0682/16 of the VEGA Grant Agency of the Slovak Republic.

REFERENCES

Bandler, W., Kohout, L. 1980. Fuzzy power sets and fuzzy implication operators, *Fuzzy Sets and Systems* 4: 13–30.

Ďuračiová, R. 2014. Implementation of the selected principles of the fuzzy set theory into spatial database system and GIS. In: *SGEM 2014, 14th GeoConference on Informatics, Geoinformatics and Remote Sensing, Albena, Bulgaria, 17–26 June, Conference Proceedings Volume I.* Sofia: STEF92 Technology Ltd.

Ďuračiová, R., Faixová-Chalachanová, J. 2018. Fuzzy Spatio-Temporal Querying the PostgreSQL/Post-GIS Database for Multiple Criteria Decision Making. In: *Dynamics in GIScience, Lecture Notes in Geoinformation and Cartography, Springer International Publishing AG*: 81–97.

Giles, R. 1976. Lukasiewicz logic and fuzzy set theory, *Internat. J. Man-Machine Stud.* 8: 313–327.

Grabisch, M., Marichal, J.-L., Mesiar, R. and Pap, E. 2009. *Aggregation Functions. Encyclopedia of Mathematics and its Applications, No. 127.* Cambridge: Cambridge University Press, Cambridge.

Gupta, M.M., Qi, J. 1991. Theory of T-norms and fuzzy inference methods. *Fuzzy Sets and Systems* 40: 431–450.

Kacprzcyk, J., Petry, F.-E., Yazici, A. (2010) *Uncertainty Approaches for Spatial Data Modeling and Processing: A decision Support Perspective*, New York: Springer Science and Buissnes Media LLC.

Klir, G.J. and Yuan, B. 1995. *Fuzzy Sets and Fuzzy Logic: Theory and Application.* New Jersay: Prentice-Hall PTR.

Lieskovský, T., Ďuračiová, R., Karell, L. 2013. Selected mathematical principles of archaeological predictive models creation and validation in the GIS environment. *Interdisciplinaria Archaeologica—Natural Sciences in Archaeology*, IV (2): 33–46.

Lieskovský, T., Faixová Chalachanová, J., Ďuračiová, R., Blažová, E., Karell, L. 2015. Archeologické predikčné modelovanie z pohľadu geoinformatiky. Metódy a princípy. 2. preprac. vyd. Bratislava: Slovenská technická univerzita v Bratislave, CD-ROM, 246 p.

Menger, K. 1942. Statistical metrics, *Proc. Nat. Acad. Sci. U.S.A.* 28: 535–537.

Petry, F.E., Robinson, V.B., Cobb, M.A. 2005. *Fuzzy Modeling with Spatial Information for Geographic Problems.* Berlin, Heidelberg: Springer-Verlag.

Schweizer, B., Sklar, A. 1983. *Probabilistic Metric Spaces.* Amsterdam: North-Holland.

Shekhar, S., Xiong, H. 2008. *Encyclopedia of GIS.* New York: Springer.

Shi, W. 2010. *Principles of modeling of uncertainties in spatial data and analyses.* Boca Raton: CRC Press/Taylor & Francis.

Shi, W., Wu, B., Stein, A. (eds.) 2016. *Uncertainty Modelling and Quality Control for Spatial Data.* Boca Raton: CRC Press.

Weber, S. 1983. A general concept of fuzzy connectives, negations and implications based on t-norms and t-conorms, *Fuzzy Sets and Systems* 11: 115–134.

Zadeh, L. 1965. Fuzzy Sets. *Information and Control* 8: 338–353.

Zimmermann, H.J. 2001. *Fuzzy Set Theory—and Its Applications. Fourth Edition.* New York: Springer Science and Business Media, LLC.

Advances and Trends in Geodesy, Cartography and Geoinformatics – Molčíková et al. (Eds)
© 2018 Taylor & Francis Group, London, ISBN 978-1-138-58489-1

SQL base for managers and visitors of the Kielce botanical garden

Ryszard Florek Paszkowski

Department of Geomatics, Faculty of Environmental, Geomatic and Energy Engineering,
Kielce University of Technology, Kielce, Poland

ABSTRACT: The Kielce botanical garden was established in 2004 in Karczowka Hill area, Kielce, Poland. On the formal cooperation contract between Kielce University of Technology and GEOPARK, Municipality, Kielce, since 2014 our B.Sc. students were involved in surveys for inventory of Kielce Botanical garden. Since 2014 till now 17 B.Sc. students were surveying and documenting the garden infrastructure elements and plants. The GNSS-RTK and statistical methods were applied. For single plants and their details, like leafs, fruits, needles, photogrammetry was applied. Totally 12.000 plants were surveyed and documented individually and in groups.

1 INTRODUCTION

1.1 Area of interest

The GNSS RTK and photogrammetric surveys of the Kielce Botanical Garden were applied for an inventory and creation of SQL data base with contents of all plants. SQL base as a *Structured Query Language* base enables obtain answers for multilevel questions concerning all data base content, [Czajkowska E, Socha K., 2014], [Czajkowska E, Łuszczyński J., 2015].

The intensive development of industry, agriculture and mining on our planet caused serious limitation and sometimes even liquidation of rare plants in many countries and world arias.

The botanical gardens were established mainly for educational purposes but also to preserve rare plants. Only in Poland 41 botanical gardens exist in major cities. The Kielce Botanical Garden was established in 2004 in area of Karczowka Hill of 339 m altitude on 15 ha area– 12 ha on a southern part and 3 ha on a northern part (Fig. 1). In 2016 a total number of plants within the garden reached over 12.000. It is too many plants to manage them manually without sophisticated computer assisted tools like interactive and intelligent SQL data base.

We should mention also about arboretums which are established mainly for scientific research programs and development of rare plants.

Figure 1. The main part of Kielce Botanical Garden on a orthophotomap as a reference layer.

Therefore, two main purposes were defined to achieve concerning the garden matters. Firstly, to help to the staff responsible for management of the garden and secondly, to give some easy tools for visitors to plan a tour through the garden not only personally but virtually as well.

1.2 *Requirements of managers and visitors of botanical gardens*

Each botanical garden needs a proper and modern management what gives opportunity to arrange a friendly access for visitors to all plants and garden infrastructure. In 2016 a total number of plants within the Kielce Botanical Garden reached over 12.000. It is too many plants to manage them manually without sophisticated computer assisted tools like interactive and intelligent SQL data base. There are many different activities of the garden managers which require almost day by day assistance what should be done at every week. The main such activities concern regular maintenance and treatment of plants like cuts, fertilizing, chemical protection, cold protection, and others. The solution of that problem is to establish sophisticated SQL data base of plants which will be very helpful tool for managers by guiding them about all required actions.

On the other hand, all information about the garden plants and infrastructure will be very useful to arrange on-spot or virtual access to the garden assets for visitors. So today, we have even options to visit the botanical gardens personally or remotely through Internet as well. Especially, it is fantastic possibility to visit the botanical gardens not only fare from our place in our country but also abroad and—on other continents, thousands kilometers away. No necessary money for travel and accommodation.

I spent some beautiful years in Cape Town, South Africa with the Geomatic Department of University of Cape Town. Now I am with the Kielce University of Technology and my computer startup screen is full of "Kirstenbosch Botanical Garden" flowers, remembering me so happy days of my life while I was walking through this garden routes under South African sunny sky—just remembrance but so beautiful (Fig. 2), [Kirstenbosch sanbi 2017].

The Kielce Botanical Garden is part of GEOPARK institution which belongs to the Kielce Municipality. A cooperation contract was signed between GEOPARK and the University and on that base since 2014 our B.Sc. students were involved in surveys for inventory of Kielce Botanical Garden as a part of their B.Sc. thesis under my supervision as thesis promotor. The theses were prepared mainly by two student teams as all required their involvement in surveys with use GNSS-RTK method and photogrammetry [Florek-Paszkowski, 2010].

Whole garden was divided on thematic sectors characteristic for kind of plants—bushes and trees. Some of plants, especially small, were located in groups. Totally 17 students were surveying and documenting the garden infrastructure and plants. Presently only part C is developed with road infrastructure and dedicated plants (Fig. 3).

Students were involved in measurements using the Kielce University of Technology GNSS and photogrammetric equipment. All works for the garden were made in close cooperation with the garden management staff, especially director Katarzyna Socha (Fig. 4).

Figure 2. The map and view of the Kirstenbosch National Botanical Garden located in vicinity of Table Mountains.

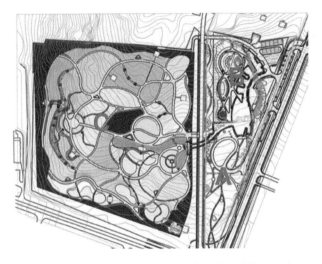

Figure 3.   The Kielce Botanical Garden parts A, B, C dedicated to different plants.

Figure 4.   Part of the Kielce Botanical Garden during inspection by the garden director Katarzyna Socha, M.Sc.

2   SURVEYS WITH USE GNSS AND PHOTOGRAMMETRY

Measurements of infrastructure and some plants were made by Global Navigation Satellite System—RTK method. All roads and pathways were measured with RTK accuracy of 2–3 cm as a mean error.

Photogrammetric documentation survey was applied especially for plants and their details like flowers, fruits and leafs. All documented plants and infrastructure elements were mapped on an orthophotomap as a reference layer.

Details of plants like flowers, fruits and leafs were also documented by surveying a location using GNSS-RTK, tachymetry and photogrammetry (Figs. 5–7).

3   BASE MAP AND ORTHOPHOTOMAPS AS A REFERENCE LAYERS
   TO MARK DOCUMENTED PLANTS AND THEIR GROUPS

Base map was established in 2015 and consist of control points, borders, height contours, infrastructure element like roads and pathways. The coordinate system 2000 and 1:500 scale were applied (Fig. 8).

Figure 5.   Part of the Kielce Botanical Garden infrastructure like road and pathways for visitors.

Figure 6.   a) on left—Korean fir (in Latin—Pinus koraiensis) – sample documentation photo of whole tree, b) on right—sample of pathway.

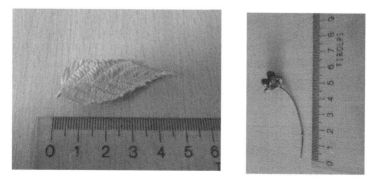

Figure 7.   On left—Leaf and on right—fruit of four petal (in Latin—Rhodotypos scandens) – a sample documentation photo of detail.

162

Figure 8. Part of base map in scale 1:500 made by Marchewka Aleksandra.

Figure 9. Part of orthophotomap 1:2000 with collections of chosen groups of plants marked in different colors.

Table 1. The structure of established SQL base consists of the following components.

Plant ID number	Polish name	Latin name
Plant type	Age since planting	Planting date
Trunk dimension	Country of origin	Place of origin
Planting person	Location sector and coordinates	Necessary maintenance
Plant diseases	Occasional planting	Reminders
Remarks	Plant family	Plant leafs or needles
Plant fruits	Total number of plants	Plant category
Plant requirements	Vermins and pests for plant	Plant height
Cold and frost resistance of plant	Plant flourishing time	Plant smell
Photo of plant	Photo of plant details	Maintenance activity details and time
Plant health status		

As for many different applications an orthophotomap is very useful reference layer for infrastructure elements like roads, pathways, constructions and single plants like trees and their groups as well (Fig. 9).

4 SQL DATA BASE FOR DATA STORAGE AND PROCESSING— STRUCTURE AND CONTENTS

All above 31 components of data base (Table 1) were discussed with the garden management staff, mainly with the garden director, Katarzyna Socha, M.Sc. Initially the base was established as the four independent thematic bases but finally we have decided to aggregate them to one SQL base for much easier query of data records. In 2017 some cadastral aspect will be added to the base like relevant parcel numbers and sector borders with reference to cadastral parcels.

5 SUMMARY

The GNSS-RTK and statical methods were applied. For single plants and their details, like leafs, fruits, needles, photogrammetry was applied. Totally 12.000 plants were surveyed and documented individually and in groups. It has been significant contribution of our University to the Kielce Botanical Garden.

A base map in scale 1:500 was made and orthophotomap was applied as a reference layer. For management purposes and visitors requirements the SQL data base of plants and group of plants was setup. The base consists of 31 components which can be used for management and guiding visitors as well. There is a pending project about on-line interactive information table at the entrance to the garden for visitors giving access to SQL data about plants and infrastructure. It should be also accessible through internet.

6 RECOMMENDATIONS AND CONCLUSIONS

1. All surveying and positioning works should be performed continuously using GNSS and photogrammetry as very applicable methods. Orthophotomap layer is the most understandable form of map giving us easy orientation almost for everybody so is very applicable and helpful for visitors.
2. There is strong need to establish the SQL data base as a very suitable tool for storage and processing acquired data about plants and infrastructure. As of huge number of records it is usually impossible to manage such data manually and it require a computer support with SQL base.
3. The management activities like cuts, fertilizing, chemical protection, cold and frost protection, watering should be guided week by week with use of SQL data base support.
4. Data of SQL base should be accessible for physical and virtual visitors of different age in different forms especially descriptions of plants, their photos, galleries, panoramas, audio guides, educational routes.

REFERENCES

Czajkowska, E., Socha, K. 2014. Kielecki Ogród Botaniczny w budowie. Kielce Botanical Garden under development. GEOPARK, Kielce, Poland.
Czajkowska Elżbieta, Łuszczyński Janusz, Kielecki Ogród Botaniczny—Przewodnik, The Kielce Botanical Garden—Guide, 2015, Geopark, Kielce.
Florek-Paszkowski R., 2010. Ortofotomapa i jej techniczne, formalne i prawne aspekty przydatności oraz wykorzystania w katastrze i gospodarce nieruchomościami. *Orthophotomap and its technical, formal and legal aspects of usefulness and applicability in cadaster and real estate economy.* XVII All Poland Symposium—Modern methods data capture and modelling in photogrammetry and remote sensing. Wrocław, Poland.
Kirstenbosch SANBI, The most beautiful garden in Africa, 2017, from sanbi.org.

Advances and Trends in Geodesy, Cartography and Geoinformatics – Molčíková et al. (Eds)
© *2018 Taylor & Francis Group, London, ISBN 978-1-138-58489-1*

Displaying of easements in vector cadastral maps

Ľ. Hudecová & R. Geisse
Faculty of Civil Engineering, Slovak University of Technology, Bratislava, Slovakia

ABSTRACT: Easements that are bound to a part of the plot are being displayed in a separate layer of vector cadastral map since 2009. Their number is gradually rising and demonstrating that the current display system is inadequate.

New guidelines for displaying of easements in vector cadastral map are a result of practical experience and analysis of current issues e.g. overlapping line objects and area objects, lack of transparency, lacking interconnection to file of descriptive information. Our solution addresses list of map symbols, list of colors, predefined object attributes, display system for servient tenement. Specifically, our solution addresses methods and means of cartographic expression of easements for institutions and individuals with interactive access rights to vector maps and those with access rights through web portal.

1 INTRODUCTION

Easements legally restrict owners of real estate in favor of another party. Restrictions include obligations to bear and withhold or call to action (Civil Code No. 40/1964). Burdened real estate should provide wholesome utilization and other necessities in favor of legally concerned entity (Mika & Leń 2016). The earliest currently registered easements in cadaster of real estates were recorded in the Austro-Hungarian Monarchy era. From the year of 1853 they were entered into land register, after the year of 1964 they were entered into owner's folio. If the easement was bound to a part of the plot, the survey sketch was attached to the legal document for registration. The extent of easement was not displayed in a cadastral map; it was only displayed in survey sketch.

In year 2009 easements that are bound to (only) a part of the plot (only "easements" so forth), began to be displayed also in the vector cadastral maps (VCMs) (Decree No. 461/2009).

VCMs have standardized digital form. Its contents are saved in files with prescribed structure on a memory drive of a computer. Contents format is suitable for processing in programs supporting CAD/CAM. Graphical elements of VCM (points, lines, areas) are grouped into objects that can have text information from description information file attached via attributes. The objects of VCMs are grouped into thematically organized layers (Kusendová 2009).

VCMs from the entire area of Slovakia are publicly available at web portal free of charge (ZBGIS Map Client application—the Real Estate Cadaster). Web portal enables browsing of cadaster of real estate maps and additional spatial data (such as ortophotographs).

Number of easements is gradually growing and demonstrating that the current display system is inadequate and convoluted (Gašincová et al. 2016), (Polaufová & Katona 2016), (Kotka & Chromčák 2016). The proposed solution differentiates methods and means of cartographic expression of easements for vector maps and web portal. Solution is beneficial for citizens and anyone who does not have interactive map access. Various map symbols, methods and means of cartographic expressions are used.

2 ANALYSIS OF CURRENT STATE OF EASEMENTS DISPLAY

2.1 *Technological and legislative aspects (2009–2016)*

In the past the burden of easements was not displayed in cadastral maps but only in survey sketches. Maps were made of paper and the display of easements would decrease their transparency and legibility.

Gradual transformation of cadastral maps into vector form removed the limitations for displaying of easements (Hudecová 2011). In year 2009 a separately themed layer "TARCHY" ("burdens" in English) for VCMs was introduced (Trembecká et al. 2010). This change was not retroactive; only newly registered easements were added after the layer creation. Every easement was displayed using a standardized line "0.141" (STN 013411) (lines used in cadaster of real estate are defined by Slovak Technical Standard).

Shortcoming of this solution was lack of informative content in the TARCHY layer. Lines were not properly specified and did not contain any attributes. It was impossible to distinguish easement category (line/area, with/without servient tenement ...), paired lines, and identify overlapping lines or link the lines with description file of cadastral documentation.

VCMs only provided user with one information; that being, whether there is an easement burdening concerned real estate. For more information about the easement it was necessary to examine the survey sketch, owner's folio and public documents concerning the entry of easement into cadaster of real estate. This state of easements display was proven to be unsatisfactory.

2.2 *Technological and legislative aspects (from 2017)*

From the year 2017 a new technical guideline for VCMs is in place (Guidelines 2016):

- all easements are displayed in VCMs (layer TARCHY) (applying retroactively);
- if there is a servient tenement applicable to concerned easement, it is required to be also displayed in VCMs;
- one object in TARCHY layer represents an easement defined by one public document;
- every object includes attribute "E" (cadastral proceeding) which links the line with description file of cadastral documentation (Fig. 1);
- scope of easements includes these aspects:
  a. area (such as servient tenement or the right to cross a foreign real estate), displayed by dot-dashed line "0.141" (STN 013411);
  b. line (such as mains axis), displayed by dot-dashed line 0.091 (STN 013411);
  c. area and simultaneously line (such as mains and its servient tenement) displayed by combination of lines "0.141" and "0.091", (Fig. 1).

Solution that was applied in 1.1.2017 introduces benefits for those who can interactively operate VCMs (directly in graphical system). Surveyor or cadastral officer can identify whether the line is part of a line, area or combined object as well as distinguish paired lines

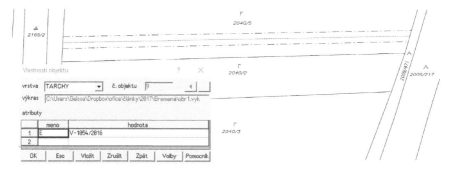

Figure 1.   Displaying of easement in VCM since 2017.

(such as servient tenement) and overlapping lines. Attribute "E" (cadastral proceeding) enables immediate identification of connection with description file of cadastral documentation.

Regular citizen can see **TARCHY** layer through web portal displayed in purple color (ZBGIS Map Client application—the Real Estate Cadaster). Other information regarding easements is not accessible to citizens.

3 METHODS AND MEANS OF CARTOGRAPHIC EXPRESSION OF EASEMENTS

One object can be interpreted through various map symbols, methods and means of cartographic expressions. The choice of tools depends on several factors; such as map form and purpose.

3.1 Line symbols method

Line symbols expression method is one of the most used methods. Easements are represented via area (in cases such as servient tenement or the right to cross a foreign real estate), single line (mains axis) or multiple lines creating a belt. In cases of four easement categories (EC) three outlines on map which define heterogeneous areas and one identifying line were proposed, Table 1.

3.2 List of map symbols

The choice of line width, structure and color is virtually unlimited. Thematic map administrators use custom highly detailed lists of map symbols and lists of colors to express specific properties. In case of mains lines a national agreement is in place and list of symbols is published in Slovak Technical Standard (STN 013411), Table 2.

3.3 Area symbols method and list of colors

Area symbols expression method is used to highlight and distinguish areas. Qualitative distinction among areas can be achieved through color (tone, shade, saturation). Table 3 is proposed use of color tone for servient tenements energy lines.

Table 1. Easement categories.

Easement Category (EC)		Line symbol	(STN 013411)	Symbol
1	area	outline	0.141	.—.—.—.—.—.—.
2	mains axis	line	0.091	....
3	belt	outline	0.031	— — — — — —
4	planned belt	outline	0.161	—..—..—..—..

Table 2. List of symbols for mains lines (selection).

Easement Lines Type (ELT)		Symbol	STN 013411
1	Energy lines	— — ∿ — — — — — — —	6.5920
2	Gas lines	—— ⌐⌐ — —— —	6.3220
3	Oil pipelines	—— ● — —— —	6.8020
4	Thermal lines	—— + — —— —	6.5120
5	Water lines	—— → — —— —	6.1520
6	Sewage lines	——)— — —— —	6.2320
7	Electronic communication lines	— — ∼ — — — — — —	6.7020
8	Cable collector	—— ═ — —— —	6.8120
9	Other lines	— — — — — — — — —	0.041

Table 3. Energy lines with servient tenements (selection).

Energy lines Easement Servient Tenements (EST)	Servient tenement	Area color [m]
1 Overhead lines (1 kV–35 kV), isolated conductors	4	
2 Overhead lines (35 kV–110 kV), isolated conductors	15	
3 Overhead lines (110 kV–220 kV)	20	
4 Overhead lines over 400 kV	35	

Almost every type of mains line is regulated through dedicated legal standard. Apart from legally defined size of servient tenement transmission companies define specific rules of protection (such as prohibition to plant and grow permanent crops, plant trees, build buildings, fences, dig wells).

3.4   Vector map attributes and interactive cartography methods

In case of vector map form factor the included information content is expanded through use of attributes. Attributes are used in graphical systems for purposes of object selection and in cases when the map is included in CAD system also for linking with database (chap. 2.2) (Kusendová 2009).

Web based interactive map can have hypertext structure which makes it generally superior to analog or digital maps. Apart from map symbols it can utilize clickable layers and hyperlinks.

4   PRINCIPLES OF METHODOLOGY

Proposed easement displaying includes separate solutions for VCMs and web portal. This approach warrants several advantages. Solution is beneficial for citizens and anyone who does not have interactive map access (public administration institutions and commercial subjects). Another benefit is the small scope of proposed changes in displaying of easements in VCMs. Adjustments and additions to the current rules are only minor (chap. 2.2).

Proposal for VCMs:

– Amend TARCHY layer by adding missing information about the range of restrictions (use line symbols and prefer attributes over other expression means).

Proposals for web portal:

– Create a comprehensive interactive map with hypertext structure (hyperlink map to description information file),
– Introduce clickable objects in map layer TARCHY, use attributes to select objects, hyperlink attributes to electronic collection of laws on web portal Slov-lex (Slov-lex). Slov-Lex provides the consolidated and up-to-date wording of all types of legislation.
– Display easements using a user-friendly list of map symbols, color symbols and other methods and means of cartographic expressions.

5   RESULTS

Proposed displaying for easement "the right to cross a foreign real estate" in VCM and on web portal is in Figures 2–3 and for easement "energy line with servient tenement" is in Figures 4–5.

Solution for VCMs retains attribute "E" (cadastral proceeding), adds mandatory attribute "EC" (easement category) (Table 1), for mains lines mandatory attribute "ELT" (easement lines type) (Table 2) and facultative attribute "EST" (servient tenement) (Table 3). Apart from two currently used lines "0.141" and "0.091" (chap. 2.2) the solution adds two additional lines "0.031" and "0.161" (Table 1).

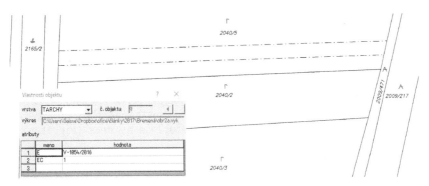

Figure 2. Displaying of easement "The right to cross a foreign real estate" in VCM.

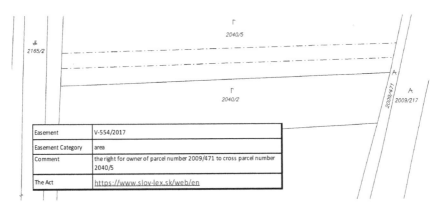

Figure 3. Displaying of easement "The right to cross a foreign real estate" in web portal.

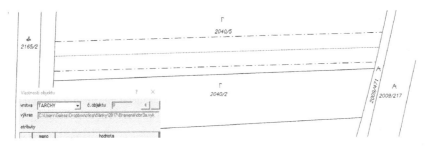

Figure 4. Displaying of easement "Energy line with servient tenement" in VCM.

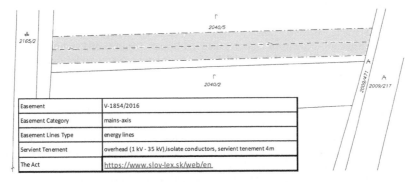

Figure 5. Displaying of easement "Energy line with servient tenement" in web portal.

Web portal utilizes a wider range of methods and means of cartographic expressions of easements, e.g. areas are colored (Fig. 5). Map application is also extended with a new interactive clickable layer (Figs. 3, 5):

- Explains contents of attributes E, EC, ELT a EST, such as "easement category = area" (Fig. 3), "easement category = mains-axis", "easement lines type = energy lines", "servient tenement = overhead (1 kV–35 kV), isolated conductors, servient tenement 4 m" (Fig. 5).
- Links the option to open respective law that is affecting the easement, e.g. "Act = Act No. 40/1964 Civil Code" (Fig. 3), "Act = Act No. 251/2012 Energy lines" (Fig. 5).

6 SUMMARY AND CONCLUSIONS

VCM is a technical base for real estate registration. Its content, form factor and displaying of easements in TARCHY layer is therefore required to fully accommodate this purpose. Thus methods and means of cartographic expression of easements in VCMs and in web portal are distinct.

Discerning of individual types of easements was achieved through detailed categorization (Table 1). Apart from list of map symbols (Table 2) the solution also contains predefined object attributes and display system for servient tenement (land burdened by easement) (Table 3). Proposed adjustments to displaying of easements in VCMs are only minor. However, functionality of web portal would be considerably expanded. Proposed visualization of easements via interactive clickable layer provides citizens, public administration institutions and commercial subjects with transparent and comprehensible information. Hyperlink to electronic collection of laws on web portal Slov-lex will provide access to detailed information about servient tenement and rules of protection. Our solution is applied and explained using a case involving the right to cross a foreign real estate (Figs. 2–3) and servient tenement around energy lines (Figs. 4–5).

REFERENCES

Act No. 40/1964 Civil Code.
Decree No. 461/2009 of the Office of Geodesy, Cartography and Cadaster of the Slovak Republic, implementing The Act No. 162/1995 on the real estate cadaster and the entries of ownership and other rights to the real estates (the cadaster act).
Gašincova, S, Weiss, G., Gašinec, J. & Labant, S. 2016. Structure, quality and methods for updating a vector cadastral map registered in the cadastral documentation in the Slovak republic. In 16th International Multidisciplinary Scientific GeoConference SGEM 2016, *Conference proceedings book 2, vol. 2*: pp. 547–554. Bulgaria: Albena.
Guidelines for recording of the scope of easements in file of geodetic information of cadaster of real estates. Office of Geodesy, Cartography and Cadaster of the Slovak Republic. 2016.
https://www.slov-lex.sk/web/en.
Hudecová, Ľ. 2011. Vector Cadastral Maps. *Geodetický a kartografický obzor* 57 (99): 243–248.
Koťka, V. & Chromčák, J. 2016. Actualization of large-scale maps symbols. In 16th International Multidisciplinary Scientific GeoConference SGEM 2016, Conference proceedings book 2, vol. 2: pp. 47–51. Bulgaria: Albena.
Kusendová, D. 2009. Application Tools of Geographic Information Systems in Demogeography. In Russian and Slovakia, *Modern Tendencies of Demographic and Socioeconomic Processes*: 158–180. Ekaterinburg, Russian Academy of Sciences the Ural Branch Institute of Economics.
Mika, M. & Leń, P. 2016. The inventory of databases on the land registration for the ecological sites in Krakow. *Ecological Engineering* 50: 121–131.
Poldaufová, O. & Katona, P. 2016. Easements with a public-law element. *Slovenský geodet a kartograf* 21(2): 9–11.
Slov-Lex Legislative and Information Portal. Ministry of Justice of Slovak Republic.
STN 01 3411 Map drawing and symbols. (Slovak Technical Standard). Slovak Standards Institute.
Trembeczká, E., Gašincová, S. & Gašinec, J. 2010. Amending act No. 162/1995 on the real estate cadastre and the entries of ownership and other rights to the real estates (the cadaster act) effective from 1.09.2009. *Acta Montanistica Slovaca* 15(1): 86–94.
ZBGIS Map Client application—the Real Estate Cadaster theme. Office of Geodesy, Cartography and Cadastre of Slovak Republic. https://zbgis.skgeodesy.sk/mkzbgis/sk/kataster.

3D visualization of extraction progress in the quarry Lehôtka pod Brehmi using tools of digital cartography

V. Hurčíková, S. Molčíková & T. Hurčík
Faculty of Mining, Ecology, Process Control and Geotechnologies, Technical University of Košice, Košice, Slovakia

ABSTRACT: This paper presents the possibilities of using modern methods of spatial visualization in surface mining. Currently, as a documentation according to official regulations are used the mine maps. Their disadvantage is that these are static two-dimensional maps to show the status in the certain point in time. In our research, we focused on the use of modern means of expression of dynamic computer cartography applied to visualize of progress of mining works in quarry Lehôtka pod Brehmi. Our goal was to create a 3D model of locality, including geology, based on available mine surveying documentation.

1 INTRODUCTION

The progress in mining works in a quarry is usually visualized by standard static maps. In this case, the basic mine map constitutes the static map. The basic mine map is a part of mine surveying documentation and is created and maintained in accordance with the generally binding legal regulations in force, represented by mine surveying regulations. Mine surveying regulation is a regulation of the Ministry of Economy of the Slovak republic governing the creation, management, refilling and maintaining of the mine surveying documentation in mining activities and certain activities carried out using a mining method (Mine surveying regulations, 1993).

The rapid development of computer and GIS technologies allows for the use of more sophisticated and more representative visualization methods at present, such as creating 3D models, the use of tools for creating dynamic maps, creating animations and the like.

As a part of the GIS applications, computer cartography offers, besides conventional cartographic methods that enable to create static maps, new ways of visualization of geographic data as well. These so-called digital "cartoproducts" include among others dynamic, multimedia and interactive maps and atlases, map servers, and virtual scenes. Dynamic maps express change of status of the mapped phenomenon in the course of time (Döllner and Kersting, 2000; Kersting and Döllner, 2002).

Three main ways of representation of temporal changes can be differentiated: a simple static map (the change is expressed, e.g. in colour differentiation), a series of static maps (for every epoch a simple static map is created and changes in the phenomenon can be seen on the chronologically successive maps) and animated maps (the change is viewed in a frame in which all maps are displayed consecutively) (Kraak and Ormeling, 2005).

It is cartographic animation that has substantiation in the visualization of spatially and temporally variable phenomena for its illustration, because it provides an absolutely different insight into the displayed data.

Cartographic animation in the context of scientific research is used for research, verification or presentation of the displayed data. Animation explains the process, but it also has the ability to clarify the relationship and trends that are not clear from the static maps (Jung, 2009; Kraak and Ormeling, 2005).

It is possible to animate the changes of one layer only or, if necessary, also the changes of multiple layers simultaneously. These changes can be recorded at different time intervals (days, weeks, years), or they can be intensive and presented slowly in the animation and that is why these changes can be captured which would not otherwise be difficult.

A distinction can be drawn between two types of animation: presentation and interactive. Presentation animation cannot be interfered with, or alternatively it allows minimal interference, hence, it can be viewed and monitored only. Interactive animations can be entered and some of its parameters can be adjusted (to zoom the view, to rotate the map, etc.).

2 LOCALITY

The quarry of Lehôtka pod Brehmi was chosen as the object of research. The quarry of Lehôtka pod Brehmi is located in the central part of Slovakia near the village of Lehôtka pod Brehmi in the district of Žiar nad Hronom. It belongs to the group of central Slovakia volcanic field on the outskirts of the Štiavnica Mountains in conjunction with the Žiar Basin.

Perlite is excavated in the quarry. Quarry Lehôtka pod Brehmi is one place only where perlite is mined and processed in Slovakia (Blistan et al., 2016).

Regular measurements of the progress in mining works are carried out in the quarry for the purposes of drawing up the basic mine map of the quarry. The measurements of changes are done once a year, at present using the GNSS technology.

The model creation was based on the measurements made during the period from 2003 to 2013 and on their grounds the mine maps were drawn up. Individual stages of measurement were processed using the software Kokeš and the final maps were also drawn up in this environment. Other sources were the geological sections of the locality which were available in analogue form (paper maps).

3 PREPARATION OF DOCUMENTS FOR THE MODEL CREATION

Prior to the creation of the model quarry, it was necessary to solve several problems related to the source data. Not only was it important to work out the compatibility of the outputs from all processing software, but to ensure the digitization of analogue source materials as well (Caumon et al., 2009; Bonaventura, 2017).

The final visualization of progress in mining works in the quarry was conducted using the software ESRI ArcGIS Desktop 10.3, therefore, it was necessary to process all the source materials to make them fully compatible with that software.

3.1 *The processing of available documents into a format supported by ArcGIS*

As mentioned before, there were mine maps available in the filename extension format .vyk in the program Kokeš. This file format was not ArcGIS supported, thus, the maps were converted to a format .dgn (design) which was the basic drawing format in the products of the company Bentley Systems and it was easily integrated into the ArcGis environment. Similarly, these were standard 2D maps based on which 3D model of locality could not be created. It was, therefore, necessary to carry the export coordinates into the program Kokeš to a file format .stx that represented the coordinates of all points and objects on the map in each year from 2003 to 2013. Points that were related to the annual changes in the quarry could not be exported, so, all files of coordinates had to be further edited and modified to include the points that described the changes within the quarry. The files of coordinates prepared this way could then be imported into the ArcGIS environment and further processed.

The next step was to create layers which would represent break lines (edge of wall, foot of slope, various visible changes in the terrain), which were important for proper 3D model creation in the respective year. The layers of break lines were created by vectorization of source maps in the file format .dgn in the ArcGis environment.

3.2 Digitalization of the geological sections

The next phase required the geological sections be digitalized. They were available for each section in the form of analogue maps (Fig. 1).

Transformation of analogue maps into the digital form was performed by scanning. After scanning the basic maps, it was possible to digitize each section by vectorization of the individual sections. Vectorization was performed in the CAD platform MicroStation V8. The respective layers were created that represented the surface, geological sections and rocks. The fills were assigned to layers displaying rocks in such a way so that different types of rocks could be identified (Fig. 2). Afterwards, the sections were then uploaded into the 3D design, where they were registered in the coordinate system UTCN and "picked" into space after rotating them by 90°.

Geological sections created this way were then used to create the 3D model of the quarry of Lehôtka pod Brehmi.

The result of source data processing specified above was the splitting of individual elements into separate layers. The list of created layers indicating the name, the type of layer, and the description of each layer is shown in Table 1.

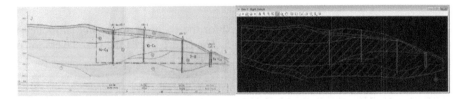

Figure 1. Geological section in the analogue form and in the Microstation environment.

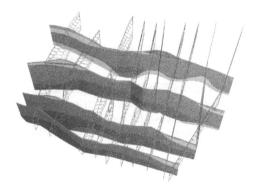

Figure 2. Geological section in the analogue form and in the Microstation environment.

Table 1. List of created layers.

Name of layer	Type	Description
Quarry_03 to 13	Point	Points representing the topography changes in particular years
Edges_03 to 13	Polyline	Break lines representing the edges of walls, foot of the slope, etc. in particular years
Polygon_03 to 13	Polygon	Border of mining works in particular years
Contours_74	Polyline	Contour lines, representing the initial topography before mining in year 1974
Section_1 to 19	Polygon	Geological sections
Buildings	Polygon	Technical buildings in quarry
Roads	Polyline	Roads in quarry
Other	Point	All others elements in map, as geological drills, devices in quarry, etc.

points
buildings
contours
terrain edges

Figure 3.   Layers in the ArcGis environment.

An example of the layers which are required for the creation of the spatial model of the quarry in particular years is shown in Figure 3. The figure shows the base layers for model portraying the status of the quarry in 2009.

4   MODEL OF THE QUARRY AND VISUALIZATION OF MINING WORKS PROGRESS

Using the layers described in Table 1, 3D models of the quarry were created for each year, for 1974 (prior to mining), and for the period from 2003 to 2013. The vector TIN (Triangular Irregular Network) representation was chosen for the models, since the TIN model provided a very good description of the heterogeneous environment. Examples of created TIN models are shown in Fig. 4.

TIN models consist of a network of irregular triangles. Since the network is irregular, we can increase the resolution of triangles in places which are essential to us, it means where the terrain is rugged. On the contrary, places which we are not interested in or which have a constant inclination or only small changes in height, can be significantly reduced in the density of triangles (Dong et al., 2012; Al_Fugara et al., 2016).

The created models were then displayed in the ArcScene application, which is a part of the 3D Analyst extension and allows to manage 3D data effectively, perform 3D analysis, create 3D objects and display layers using 2.5D visualization. Geological sections were displayed as well.

Figure 5 shows two out of a total of ten created models including the geological sections. It is a view displaying the status of the quarry in years 2003 and 2013. The figure evidences the obvious progress in mining works at the quarry in the course of ten years.

5   ANIMATED VISUALIZATION OF PROGRESS IN THE MINING WORKS

With regards to the fact that the visualized data relate to the same area and display the same structured values which undergo changes in time, it seemed appropriate to consider the visualization of the data by means of animation.

The visualization of the progress in mining works was created using the software ESRI ArcGIS Desktop 10.3 and its application ArcScene, which includes the tool Animation to create animation sequences. The tool distinguishes two basic methods of animation. The first type is time-dependent animation (Time Layer Animation) which can be used only for one

174

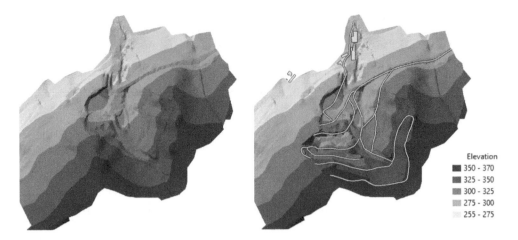

Figure 4.   TIN models of the quarry in 1974 and 2003.

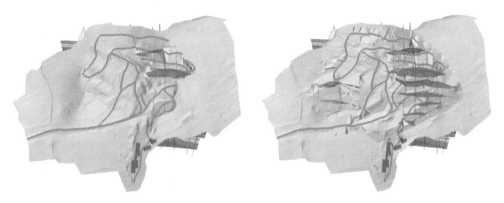

Figure 5.   TIN models of the quarry in 1974 and 2003.

layer of data with time data in the attribute table. The second type of animation that ArcGIS offers is the animation of layer groups (Group Animation). In this case, the all layers shown in ArcScene enter into animation.

After setting, the animation can be edited in two ways. The first way is via Animation Manager. The tab Track contains a list of visualized layers, through this list the properties of layers in the visualization can be approached. Another method of editing the animation is by means of the Animation Controls tool. It is possible to play the animation here on the user-configured conditions. The total time of the animation can be adjusted or the time of displaying for each of them can be set according to the number of images.

Based on the above described procedure, the animation representing the progress in mining works in the quarry of Lehôtka pod Brehmi was created. The animation was subsequently exported to .avi (Audio Video Interleaved).

6   CONCLUSION

Having processed the available archive data, a 3D model of surface quarry and of its wider surroundings was created. Moreover, the deposit area model showing the position of geological sections was created as well. The created spatial model comprehensively displays the deposit area of our interest with together with its surface status in 3D and gives a complete picture of the mutual spatial relations.

The advantage of created model of the territory is that it spatially displays the status of the quarry in respective years. At the same time, it provides the opportunity to carry out various analyses of data using the GIS tools, such as the calculation of the volume of extracted material, etc.

The model can further be expanded to a comprehensive GIS that will include all digital maps, sections, databases of realized explorations, survey measurements, databases of ore stock, etc. This GIS will then secure the uniform data management, ease to obtain data from the central database, performance of analyses, production of thematic maps and outputs.

Nowadays, dynamic computer cartography means of expression find their application in many areas, especially in natural and socio-economic sciences. The aim of the present paper was to point out the possibilities of their use in the mining industry, because they are usually not applied in this area or in the technical maps.

The standard output, showing the state of some mining area at a certain time, is, pursuant to the regulation the Base mining map. However, map enables to display the status only at a specific point in time. Unlike static map, the modern methods of digital cartography allow displaying the mapped facts as they change and evolve over the time, or as their status are in real time. One of the aims of this paper was to create a spatial model of quarry on the basis of static mining maps. The result of such processed mine surveying documentation is a 3D visualization of the quarry (Wajs, 2015; Vanneschi et al., 2014), which captures the dynamics of the progress in mining works in space and time. The created model is not of archiving nature as a Base mining map; it serves to visualize the progress in mining works in the quarry Lehôtka pod Brehmi during the period from 2003 to 2013.

REFERENCES

Al-Fugara, A'kif, Al-Adamat, R., Al-Kouri, O., Taher, S.: DSM derived stereo pair photogrammetry: Multitemporal morphometric analysis of a quarry in karst terrain. *The Egyptian Journal of Remote Sensing and Space Sciences*, Volume 19, Issue 1, pp 61–72. http://www.sciencedirect.com/science/article/pii/S1110982316300047, 2016.

Blišťan, P., Kovanič, Ľ., Zelizňaková, V., Palková, J.: Using UAV photogrammetry to document rock outcrops. *Acta Montanistica Slovaca*. Volume 21, Issue 2, pp 154–161, 2016.

Bonaventura, X., Sima, A.A., Feixas, M., Buckley, S.J., Sbert, M.: Information measures for terrain visualization. Computers & Geosciences, Volume 99, pp 9–18. 2017.

Döllner, J., Kersting, O.: Dynamic 3D Maps as Visual Interfaces for Spatio-Temporal Data. In: *Proceedings ACM GIS*, pp 115–120, 2000.

Dong, M., Hu, H., Azzam, R.: Application of three dimensional geological models to geotechnical engineering problems. In: *Rock Mechanics: Achievements and Ambitions* - Proceedings of the 2nd ISRM International Young Scholars' Symposium on Rock Mechanics, 14. – 16. October 2011, Beijing, China, pp 473–477, 2012.

John A. HowelldCaumon, G., Collon-Drouaillet, P., Le Carlier de Veslud, C., Viseur, S., Sausse, J.: Surface-Based 3D Modeling of Geological Structures. *Mathematical Geosciences*, Volume 41, Issue 8, pp 927–945. http://link.springer.com/journal/11004/41/8/page/1, 2009.

Kersting, O., Döllner, J.: Interactive 3D Visualization of Vector Data in GIS. In: *Proceeding GIS '02, Proceedings of the 10th ACM international symposium on Advances in geographic information systems*, November 08–09, 2002, McLean, Virginia, USA. http://www.nvc.cs.vt.edu/~ctlu/Project/Traffic-Project/p107-kersting.pdf, 2002.

Kraak, M., Ormeling, F.: Cartography. Visualization of spatial data. 3rd edition, Pearson Education Limited, Edingurgh, 198 p., 2005.

Mine surveying regulations: Výnos MH SR č. 1/1993 (označený v čiastke č.54/1993 Z. z.).

Vanneschi, C., Salvini, R., Massa, G., Riccucci, S., Borsani, A.: Geological 3D modelling for excavation activity in an underground marble quarry in the Apuan Alps (Italy). In: *Computers & Geosciences*, Volume 69, pp 41–54. http://www.sciencedirect.com/science/journal/00983004/69, 2014.

Wajs, J.: Research on surveying technology applied for DTM modelling and volume computation in open pit mines. Mining Science, Volume 22, pp 75–83. http://www.miningscience.pwr.edu.pl/Issue--2015,1161, 2015.

Advances and Trends in Geodesy, Cartography and Geoinformatics – Molčíková et al. (Eds)
© 2018 Taylor & Francis Group, London, ISBN 978-1-138-58489-1

Conceptual basis for a modeling cultural heritage in Slovakia: Harmonisation requirements and application the Humboldt project tools

K. Kročková

Faculty of Civil Engineering, Slovak University of Technology in Bratislava, Bratislava, Slovak Republic

ABSTRACT: An application of the data harmonisation methods to any domain working with spatial data is used approach for an elimination the data heterogeneity and assigns the data interoperability nowadays. One of them is the Cultural Heritage protection domain. In this paper we focused to a deeper analyze of proposed data specification in INSPIRE ProtectedSites theme where the Cultural Heritage protection domain should become a part. Our study was realized within the support of Humboldt project tools, work based on mapping INSPIRE ProtectedSites schema definition as an objective schema in order realize required data harmonization. As an input schema was used the dataset structure of archaeological activity register owned by the Institute of Archaeology SAS, published by web map service. The results of analyses return in general questions of the Cultural entity unit protection in INSPIRE theme. Finally stated pilot study of application the data harmonisation to archeological activity register as spatial dataset used in Slovakia on national level was taken into account to target INSPIRE ProtectedSites schema definition and gives an overview of the schema mapping and final data transformation.

1 INTRODUCTION

The Cultural heritage is composed of discrete entities which value is determined by those responsible for the research. According to agreed concept the Cultural heritage is considered to include both tangible/material and intangible/immaterial elements. Within the first category there are two basic types of elements: immovable or built, which are linked to their spatial location—a church, a bridge or an archaeological site—and movable, which can be transported without losing their main intrinsic characteristics—a painting, a piece of pottery or a sword-. On the other hand, the so-called intangible heritage comprises all those human activities whose existence is not immanent but depends on their performance, being closely linked to memories and folklore (e.g. a dance, a pilgrimage or a traditional kitchen recipe). Intangible heritage will always be linked to its spatial dimension, as the context for the human activities which generate and recreate it.

Multiple information sources with different data structure gathered in local, scholar and mainly state partial projects often duplicate the information with different degrees of complexity and efficiency. The distribution of cultural heritage information is heterogeneous with multiple cultural heritage datasets which are often fragmentary, unconnected and barely accessible. The Monuments Board of the Slovak Republic (with the regional monuments boards) as register authorities responsible for the management of cultural heritage information apply administration and give a normative framework to create and gather records on national level to support cultural heritage protection and preservation.

Application the data harmonisation in general means the possibility for an efficient and meaningful combination of heterogeneous information into consistent and unambiguous information products. Combine heterogeneous data using the conceptual data model for developing interoperability in spatial data infrastructures. This is also the main objective of the INSPIRE Directive.

As a first step towards an application of the harmonisation process of cultural heritage data is the creation of conceptual data model for used datasets. Conceptual data models are based on the relevant data covering the national register of archeological evidences in Slovakia and its complementary dataset CEANS covering archaeological activities provided by the Institute of Archaeology of the Slovak Academy of Sciences from 1945 (Bujna, J. 1993).

2 INTEGRATING CULTURAL HERITAGE INTO THE INSPIRE FRAMEWORK

European project well known as INSPIRE (Infrastructure for Spatial Information in the European Community) Directive gives standards and develops data specifications to build Spatial Data Infrastructure accordingly for Cultural Heritage information falls into Protected sites theme, part of the INSPIRE Annex I. Interdisciplinary team of specialists in the field of Geomatics and Cultural Heritage develops Cultural Heritage conceptual data model which implies adaptation to INSPIRE as well as to several ISO norms: ISO 19100 (geographical information), ISO 21127 (CIDOC-CRM Model) for heritage thematic data and ISO 15836 (Dublin Core) for document resources.

The CIDOC Conceptual Reference Model (CIDOC CRM) is well known and most used model for the development of conceptual data models for cultural heritage domain and ISO standard. Although model provides a lot of classes for localization cultural heritage objects, according to (Doerr, M., 2003) it focuses mainly to museum collections modeling and not spatial localization of cultural objects. Therefore a spatial geometry approach used in INSPIRE is more useful for cultural objects modeling.

According to (Gonzalez-Perez, C. et al., 2011) cultural heritage conceptual data model comprehends two main dimensions: cultural entities in a strict sense and the legislative figures created to protect them. This allows for the representation of cultural objects (i.e. historical buildings or archaeological sites) and their link to their legislative protection, keeping them as separate realities.

Presented Cultural Heritage conceptual data model developed by (Fernández-Freire, C. et al. 2013) is represented by The Protected Heritage Place class taken into consideration in INSPIRE ProtectedSites theme (Figure 1). It gives a definition of minimal requirements for evidence in domain and allow for each involved country admitted the INSPIRE Directive to engage to Cultural heritage protection not only on local level but in global context of domain information access. Detailed view to each feature of conceptual model and its understanding is the basement for next development in this domain in relation to assign the access to data.

According to (Fernández-Freire, C. et al. 2014) "…our model is not very prescriptive and is highly abstract, so that it may be applied to any temporal or geographical context. However, there is more detail to it than just a mere distinction between legal objects (Protected sites) and real-world objects (Cultural entities)." In proposed data model Cultural entity must be represented and registered only ones, although could be related to many Protected sites (e.g. as national monument or as UNESCO World Heritage site). This self-aggregation relationship in model is represented by function "contains" allows for this kind of behavior, enabling an aggregation of objects of the same class. The second issue, solved by the self-association function "protectionSurrounding", is the usual reference in a legal document to the site's protected surroundings, which may typically have a different legal condition and degree of protection, but whose existence is inseparably linked to the site itself. As long as the protected surroundings have a different geometry, identifier, area, etc., they should be instanced as a new object of the same class.

As a first question becomes from analysis the proposed model is missing exact definition of modeled Cultural entity in INSPIRE. A huge amount of registered information in domain gives us many kinds of information to be protected. From archaeological point of view the completely explored archeological site was already destroyed by research. There is a question;

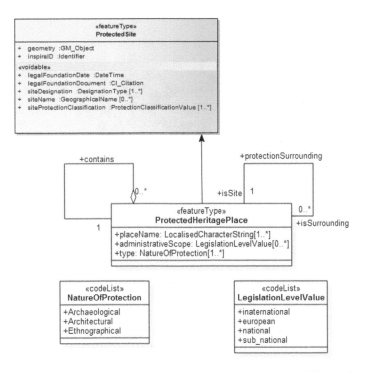

Figure 1.   The Protected Heritage Place class in relation to INSPIRE ProtectedSites application schema.

Will we model (and thus protect) archaeological resources (sites) or information from them? The only unique one and the permanent spatial unit is archaeological research, which, of course, has little in common with Cultural entities.

3   LOCALIZATION ACCURACY OF THE ARCHAEOLOGICAL RESEARCH

As was previously stated the cultural heritage information is mentioned in INSPIRE conceptual data model in the very specific form of protected sites and only a very specific part may be incorporated into that defined data specification: that of the geographical locations protected, or regulated, for their cultural value by some legal, administrative procedure. Data analysis shows that a spatial delimitation of the archaeological activities in national register dataset has not been always guaranteed, mainly for old historical records, when the requirement to a precise geographical localization was on lower level as is realized nowadays with the benefits of mapping devices development. Many records of archaeological activities miss any spatial information and on the other hand the most precise ones are delimited by spatial coordinates. All these varieties built three levels of accuracy categorization captured in mentioned national register dataset.

Levels of accuracy, how archaeological research is localized?:

1. *accurate* – identification in map; by spatial coordinates,
2. *less accurate* – identification by researcher description; by geographical site name,
3. *not localized* – identification unknown; by the reference point of cadastral area.

Dataset covers whole country area of Slovakia with 7779 point locations represented archaeological activities; listed negative as well as positive activities (Figure 2). The attribute table shows a list of attribute and some example of content in Slovak language. In attribute table is shown an example of archaeological site with three records in dataset from years 1946, 1968 and 2013 when separate archaeological researches have been done.

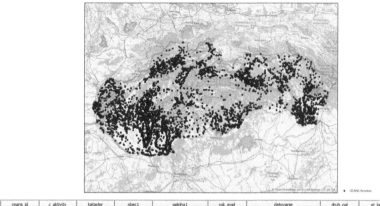

ceans_id	c_aktivity	kataster	obec1	poloha1	rok_evid	datovanie	druh_nal	st_lokal	x_jtsk	y_jtsk
1	ceans.603	Č.a. 00266	Vráble	Vráble	Fidvár (Foldvár)	1946	BZ - st, ma, RI, ST		1	-494351.90027	-1280119.22195
2	ceans.7519	Č.a. 12487	Vráble	Vráble	Foldvár (Fidvár)	1968	EN - lg, ba, kc, BZ-st,ha,ma,uh,LT-n		1	-494368.48757	-1280145.20395
3	ceans.15066	Č.a. 27246	Vráble	Vráble	Fidvár	2013	eneolit,doba bronzová		1	-494284.41827	-1280260.76844

Figure 2.   CEANS dataset: Visualization of the geometries and the content of attribute table.

Based on different values of categorization and the level of classification there is one and more dataset records belong under one cultural entity (selected three records for one object in reality) what is not required approach as mentioned in chapter 2.

4   DATA HARMONISATION

Two main phases were distinguished within a data harmonisation process: Definition of the target—INSPIRE ProtectedSite specification comprising the conceptual schema and other characteristics of examined archaeological evidence data based on well-known data content and its meaning (semantics), next step is Schema mapping—mapping between schemas from source data to match the target schema. A prerequisite for conceptual schema transformation is the definition and formalization of mappings rules for transformations between source and target schemas. This can be done on different levels of data modelling (e.g. conceptual schema or data format level) using proprietary or open mapping languages and tools. Used schema mapping according to meaning of the attributes is listed in Table 1.

The HUMBOLDT project has the aim of implementing a framework for data harmonisation in the geoinformation domain under the INSPIRE Directive. One of the main objectives of the HUMBOLDT project is to provide tools to map and transform datasets and application schemas. The HUMBOLDT Alignment Editor (HALE) tool provide user interface for defining mappings between concepts in conceptual schemas as well as for defining transformations between attributes of these schemas.

As a first step the target data schema is imported in HALE tool to establish the mapping rules for the classes and attributes of a source to the target conceptual schema. After loading our source dataset layer, we can load the target schema. The schema mapping is performed according to created matching table by selecting the appropriate mapping function. In this case we used a *Rename Attribute* function (Figure 3). We can use an *Attribute Default Value* function to fill a field with no data in the source or an *Assign* function for any code list value. Final phase is to proceed the data transformation.

All used mappings were based on a retype operation. In general, the *Retype* function states that a source and a target type are semantically the same.

To overview of transformed data we can use HALE map view with convenient cartographic visualization. On (Figure 4) you can see selected example of three archaeological research point locations of the archaeological site mentioned in previous section 3.

Table 1. The schema mapping between source and target schema.

	Dataset CEANS	Transformation process	INSPIRE target schema
Projection	**S-JTSK**, east x/north y (S-JTSK/Krovak East North) negative coordinates EPSG:5514	On the fly projection transformation	**ETRS89** EPSG:4258
Geometry	**POINT**		**POINT**
Attributes and Data types	**ceans_id**	Rename attribute	**inspireID**
	poloha1	Rename attribute	**siteName**
	rok_evid	Date extraction	**legalFoundationDate**
		Generate Unique Id	**id**
		Assign	**siteProtectionClassification** value: archaeological

In our model we set up code value with Assign function for attribute:

ProtectionClassificationValueType – value = "archaeological"

Remark: Always when a code list exists or needs to be set up new one it should be implemented through DesignationSchemeValue codelists, all expandable.

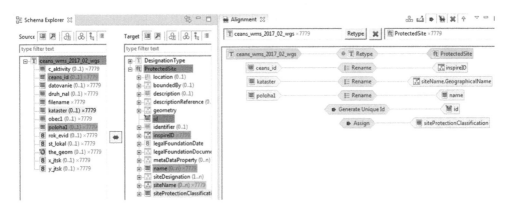

Figure 3. HALE mapping between source (CEANS dataset) and target (INSPIRE ProtectedSite) schemas.

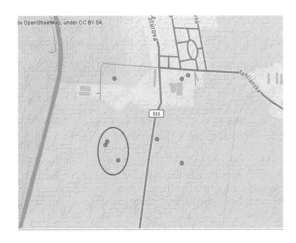

Figure 4. HALE map visualization of transformed CEANS dataset on OpenStreetMap basemap.

5 CONCLUSION

The result of the archaeological register dataset analysis within INSPIRE requirements for theme Protected sites brings a number of fundamental questions and findings a solution; what spatial unit we want to protect (to model)?, How to deal with a low accuracy of spatial units?, the fact that more than one archaeological activity record belong under one archaeological site (Cultural entity) etc. In first phase we need to agree the exact definition of protected Cultural entity unit and its properties to model and to solve the requirement where in proposed data model each Cultural entity (Protected site) must be represented and registered only ones, although could be related to more Cultural entities (Protected sites). All mentioned questions and requirements need to be further analyzed with domain experts before we start with a dataset transformation to require INSPIRE target of the data harmonization to provide access to data and to ensure interoperability in building future infrastructures.

This use-case shows a real example of mapping and transformation of national register data of archaeological evidences from the original source schema to the target INSPIRE application schema for Protected sites. Thanks to project Humboldt and his free of charge HALE tools was allowed to work on this pilot study of the data harmonization in the Cultural Heritage domain in Slovakia. With the support of domain experts we prepare the schema mapping sets based on analyzed data content and data meaning.

The proposed alignment of schema mapping like presented in this paper could be exported in many formats (XSLT, CSV etc.) and repeatedly used anytime in future without limits of new releases of dataset layer if the structure is maintained. With the release of a new database structure it is always necessary to adjust the schema mapping settings.

REFERENCES

Bujna, J., Kuzma, I., Jenis, J. 1993. CEANS—Centrálna evidencia archeologických nálezísk na Slovensku—Projekt systému. In: Slovenská archeológia XLI 2. AÚ SAV, Nitra, 1993. s. 367–386. ISSN 1335–0102.

Centrum výskumu najstarších dejín stredného Podunajska. Projekt: Archeologický ústav SAV v Nitre. [online] www.cevnad.sav.sk (12.5.2017).

Definition of the CIDOC Conceptual Reference Model, Version 5.1.2. ICOM/CIDOC Documentation Standards Group, Continued by the CIDOC CRM Special Interest Group. International council of museums. [online] http://cidoc-crm.org/get-last-official-release.

Doerr, M. 2003. The CIDOC Conceptual Reference Module. An Ontological Approach to Semantic Interoperability of Metadata. AI Magazine, 24, p. 75–92.

Drafting Team "Data Specifications". D2.5: INSPIRE Generic Conceptual Model, Version 3.4. Projekt INSPIRE, 2014. [online] https://inspire.ec.europa.eu/documents/inspire-generic-conceptual-model.

Fernández-Freire, C. et al. A Cultural Heritage Application Schema: Achieving Interoperability of Cultural Heritage Data in INSPIRE. In: International Journal of Spatial Data Infrastructures Research. 2013, Vol.8, s. 74–97. ISSN: 1725-0463 74–97.

Fernández-Freire, C. et al. A data model for Cultural Heritage within INSPIRE. CAPA 35. 2014. ISSN: 1579–5349.

Fichtinger, A. et al. (2011). Data Harmonisation Put into Practice by the HUMBOLDT Project, International Journal of Spatial Data Infrastructures Research, Vol.6, 234–260.

Gonzalez-Perez, C., Parcero-Oubiña, C. A Conceptual Model for Cultural Heritage Definition and Motivation. CAA2011 - Revive the Past: Proceedings of the 39th Conference in Computer Applications and Quantitative Methods in Archaeology, Beijing, China. 12–16 April 2011.

INSPIRE Thematic Working Group Protected sites. D2.8.I.9 INSPIRE Data Specification on Protected sites—Guidelines. Projekt INSPIRE, 2010.

Metodická pomocná inštrukcia pre vypracovanie a posudzovanie dokumentácie z archeologických výskumov, Archeologický ústav SAV, Nitra, 2017. s.5–6. [online] www.pamiatky.sk/sk/page/usmernenia-metodicke-materialy (5.10.2017).

Open structured databases' use for spatial econometrics within data-driven governance and business intelligence

B. Kršák, C. Sidor & Ľ. Štrba
Department of Geo and Mining Tourism, Faculty of Mining, Ecology, Process Control and Geotechnology, Technical University of Košice, Košice, Slovak Republic

ABSTRACT: Paper's aim is to demonstrate the power of high quality open structured data within spatial econometrics and as an essential input for building up cross-sectoral business intelligence systems' databases' fundamental content on industries' stakeholders at the lowest administrative level. The aim is explained on the practical example of linking an open structured SQL database containing all Slovak entrepreneurs' financial statements with the Eurostat methodology schema's on tourism industries stakeholders' categorization and with the Slovak Republic's Registry of destination management organizations established under the Act no. 91/2010 about the support of tourism (hereinafter R_DMO). For the purposes of linking, the schema and the R_DMO were transformed into single tables with joinable integers. Relevant queried data from the upper mentioned database operated by the "ekosystem.slovensko.digital" were geocoded using an external application running on Google Maps Geocoding API. All of the resulting data were uploaded and tested within the environment of a currently developing Destination Business Intelligence System using among other open geospatial technologies. Ultimately, the results quantify, visualize and discuss the current involvement of tourism industries stakeholders in Kosice region's (Slovak Republic) destination management organizations. Further on, the paper discusses future necessary steps that could bring more flexible solutions for both public and private stakeholders in terms of taking advantage of aggregated public data for data-driven decision making and gaining competitiveness advantage from spatial econometrics.

1 INTRODUCTION

The issue of monitoring tourism and its sustainability has been a topic for over two decades. Fuchs's et al. (2013) pointed out the fact that for a long time sustainable tourism used to be misused as a political catch phrase with several definitions, but remaining a blurred concept. Delaney and MacFeely (2014) and many other emphasize that consumption and expenditure in tourism are dispersed across a wide arc of industries: transport, accommodation, catering, entertainment, culture, sports and other related services. Not only at destination level, but in general tourism statistics have become consequently difficult to compare with other economic sectors (Delaney and MacFeely, 2014). Ionescu (2014) pointed out another issue that there are cases of inappropriate use of tourism indicators or without combinations of related indicators, thus results of analysis are put out to the danger of inconsistency. Quantifiable data has become an essential asset not only in destination management, but also in the processes of conducting evidence based policies via business intelligence processes impacting tourism (Dugas, 2015; Khouri et al., 2009; Khouri et al., 2011; McCole and Joppe, 2014).

Within the last decades and also recent years there have been initiatives and projects to bring measurement of tourism sustainability to a more tangible and practical level. In 2004, after more than a decade of research conducted by over 60 authors, the World Tourism Organization (hereinafter UNWTO) published its guidebook for "Indicators of Sustainable Development for Tourism Destinations" (UNWTO, 2004). In 2008, the United Nation's (hereinafter UN) Department of Economic and Social Affairs released the updated methodological framework for establishing Tourism Satellite Accounts (hereinafter TSA) that contains definitions, concepts and recommendations in regards to tourism statistics' observation and their analysis, mostly at national level (UN, 2008). The concept of TSA has been used also for establishing Regional Tourism Satellite Accounts (hereinafter RTSA). Examples of RTSAs may be found across European countries, but the issue of lack of availability of reliable and open data at sub national or destination level is still a critical issue. OECD's Head of the Tourism Unit warned that tourism and related data at sub national level is often weakly disaggregated, statistically invalid, not appropriately analyzed nor shared among stakeholders (Dupeyras and MacCallum, 2013). Because of lack of funding for developing statistics development and related human resources the statistical base at sub national level is still a great weakness (Dupeyras and MacCallum, 2013). It has to be pointed out that are tens of other approaches, standards. Most importantly in the recent years there have been developed a number of information and communication technology based solutions with the aim to strengthen implementation of monitoring of tourism efficiency.

In regards to sustainable tourism the European Commission (hereinafter EC) has been dealing with the issue of destinations' sustainability within its political framework for European tourism called "Europe, the world's No. 1 tourist destination". Specifically planned action no. 11 with the aim to *"Develop, on the basis of Network for Competitive and sustainable tourism regions or European Destination of Excellence, a system of indicators for the sustainable management of destinations. Based on this system, the Commission will develop a label for promoting tourist destination"* (European Commission, 2010). One of the action's results is the ETIS standard launched in 2013. The updated set of ETIS indicators is the result of several years of cooperation between the EC and Tourism Sustainability Group. ETIS's main aim as a voluntary management tool is to support destination management organizations to monitor and measure their destinations' sustainable performance (European Commission, 2016). The tourism market is an aggregation of multiple industries as a one complex cross sectoral system. Eurostat's (2014) methodology takes into account certain main economic activities (hereinafter NACE) (Table 1).

In recent years, a number of scientific approaches powered by information and communication technologies towards new ways of destination management have been developing referred to as smart destination management. Smart destinations among other, are equipped with tools capable of capturing and analyzing information for supporting efficient decision making for relevant stakeholders for the ultimate goal of improving the quality of tourists' experience (Muñoz & Sánchez, 2013). Bulhalis (2015) within smart destination management refers to the smartness narrative as taking *"advantage of interconnectivity and interoperability*

Table 1. List of tourism industry main economic activities (Source: Self-elaborated based on European Commission, 2008).

Tourism industry	NACE code	NACE name	Tourism industry	NACE code	NACE name
Main	51100	Passenger air transport	Partial	49100	Passenger rail transport
Main	55100	Hotels and simil.accomod	Partial	49320	Taxi operation
Main	55200	Holiday accommodation	Partial	49390	Oth.pass.land transp.nec
Main	55300	Camping grounds	Partial	50100	Sea passen.water transp.
Main	79110	Travel agency activities	Partial	50300	Inland passen.wat.trans.
Main	79120	Tour operator activities	Partial	56101	Eating houses
			Partial	56102	Instit.of school consume
			Partial	56109	Other purpose consume
			Partial	56300	Beverage serving activ.
			Partial	77110	Renting of cars
			Partial	77210	Renting of recreat.goods
			Partial	79900	Oth.reserv.servat.act.

of integrated technologies to reengineer processes and data in order to produce innovative services, products and procedures towards maximizing value for all stakeholders".

Destination Business Intelligence System (hereinafter DBIS) is a pilot platform under development covering Kosice region (Slovakia) and its destinations. DBIS's aim is to support decision making of public and private stakeholders of tourism via its report server's dashboard's data analytics (Kršák et al., 2015; Kršák et al., 2016a; Kršák et al., 2016b; Štrba et al., 2016).

Next lines' aim is to a) test a sample dataset's content on tourism industry active economic entities' nature within Kosice region from the perspective of share and possible employment, b) compare the results with the Kosice region DMOs' nature and c) test the outcome spatial representation with combination of open spatial data.

2 METHODOLOGY

For identifying active tourism industry entities, only not terminated entities' profiles were extracted from the slovensko.ekosystem.digital's (2017) Datahub server. For getting a better picture on Kosice region's total tourism industry, the percentage of active and terminated entities was queried. To identify the nature of the total tourism industry, grouped NACEs' entities' share were calculated. To gain more insight on NACE groups' position within the region's economy, their minimum and maximum possible employment were calculated based on the records' organization size identification tags' scaling. In addition the NACE groups' minimum and maximum share on the area's total employment were calculated with basic Structured Query Language (hereinafter SQL).

For purposes of deeper understanding Kosice region's destination management organizations (hereinafter DMOs), three DMOs membership registries' were extracted from the Datahub. Further on records were referenced in accordance with the relevant DMOs, geocoded and uploaded to DBIS's database. The spatial visualization's test with combination of was conducted within the presented paper were conducted in QGis.

3 RESULTS AND DISCUSSION

From 5 048 records on tourism industry entities registered in Kosice region, 56,00% (2 827 records) were active at the time of analysis. Details on percentage of active records may be seen be below (Table 1). Partial results indicate that most stable NACE's in terms of termination are entities with main activities *"Passenger air transport"* (100% active), "Renting of recreational goods" (73,68% active), "Hotels and similar accommodation" (73,81% active) and "Holiday accommodation" (72,67% active). The largest shares of active entities are with NACE *"Eating houses"* and *"Beverage serving activities"* (Table 1).

In terms of share on minimum and maximum possible employment, within the region's tourism industry, the largest share cover entities with NACE *"Eating houses"* and *"Beverage serving activities"*. In terms of tourism industries' share on the region's total employment with minimum of 5272 jobs and 2,72% is not so overwhelming (Table 2). From the perspective of maximum possible jobs, with 7774 jobs and 2,62%, the share is even lower (Table 2).

From the perspective of spatial representation, most tourism industry entities are located in existing DMOs' territories (Fig. 1). From the perspective of spatial representation of employment, the highest densities may also be identified the areas of existing DMOs (Fig. 2). In the south-west (Rožňava and its vicinity) and south-east parts (Trebišov, area of Tokaj Wine Region and suroundings) of the self-governing region are in terms of employment densities micro markets without an official umbrella DMO established in accordance with relevant legislation.

In terms of Kosice region DMO's memberships' coverage of tourism industry, from the total 131 entities only 25 entities were identified by their main economic activity as part of total tourism industry (overall 0,88%). Top ranking NACE groups may found below (Table 4).

Table 2. Comparison of terminated and active records in Kosice region ranked by % share of tourism industry (Source: Self-elaborated based on raw data from Slovakia.digital, 2017).

Rank	NACE code	NACE name	Total count of records	Count of active entities	Active entities %	Share on tourism industry %
1	56101	Eating houses	1216	786	64.64	27.72
2	56300	Beverage serving activities	1050	649	61.81	22.88
3	49390	Other passenger land transportation n.e.c.	934	430	46.04	15.16
4	56109	Other purpose consume	856	422	49.3	14.88
5	55200	Holiday accommodation	161	117	72.67	4.13
6	55100	Hotels and similar accommodation	126	93	73.81	3.28
7	79110	Travel agency activities	109	74	67.89	2.61
8	77110	Renting of cars	113	72	63.72	2.54
9	49320	Taxi operation	302	65	21.52	2.29
10	79120	Tour operator activities	99	65	65.66	2.29
11	79900	Other reservation service activities	55	29	52.73	1.2
12	77210	Renting of recreational goods	19	14	73.68	0.49
13	55300	Camping grounds	12	8	66.67	0.28
14	56102	Institutes of school consume	7	7	100	0.25
15	51100	Passenger air transport	5	5	100	0,18

Table 3. Ranking of active records' share on employment by minimum number of jobs (Source: Self-elaborated based on raw data from Slovakia.digital, 2017).

Rank	NACE code	NACE name	Minimum of jobs	Minimum share %	Maximum of jobs	Maximum share %
1	56101	Eating houses	1500	0.78	2286	0.77
2	56300	Beverage serving activities	1310	0.68	1989	0.67
3	56109	Other purpose consume	737	0.38	1069	0.36
4	55100	Hotels and similar accommodation	584	0.30	982	0.33
5	49390	Other passenger land transportation n.e.c.	499	0.26	572	0.19
6	49320	Taxi operation	136	0.07	209	0.07
7	55200	Holiday accommodation	116	0.06	143	0.05
8	79120	Tour operator activities	112	0.06	146	0.05
9	79110	Travel agency activities	91	0.05	117	0.04
10	77110	Renting of cars	73	0.04	84	0.03
11	56102	Institutes of school consume	40	0.02	75	0.03
12	51100	Passenger air transport	28	0.01	52	0.02
13	79900	Other reservation service activities	26	0.01	26	0.01
14	77210	Renting of recreational goods	12	0.01	12	0.00
15	55300	Camping grounds	8	0.00	12	0.00

Spatial representations of tourism industry entities participation in DMOs indicate that all three DMOs not yet reached their full potential as local main bridges between the public and private sector (Figures 3–5).

Currently there are several applications with different approaches to benefit from business intelligence (hereinafter BI) via ETL (Extract, transform, load data), OLAP (online analytical processing), data storing, data analytics and several user interfaces with all sorts of functionalities for data conversion, modeling, reporting etc. (NT, 2017). Usually the main targets of these solutions are enterprises as single end users, so the main aim is to support one user's economic growth based mainly on the end user's data. In the case of DMO's and tourism

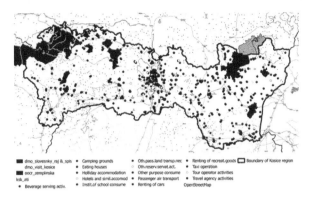

Figure 1.   Concertation of active tourism industry entities (Source: Self-elaborated in QGis).

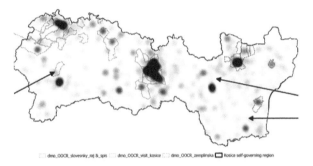

Figure 2.   Kernel density of tourism industry entities within Kosice region from the perspective of minimum employment (Source: Self-elaborated in QGis).

Table 4.   Top 10 ranked NACE groups in Kosice region's DMOs (Source: Self-elaborated).

Rank	NACE code	NACE name	Number of entities
1	84110	General publ.admin.act.	28
2	94992	Act.of hobies org.	11
3	55100	Hotels and simil.accomod	9
4	56109	Other purpose consume	6
5	55909	Accom.in stud.residences	5
6	56101	Eating houses	4
7	91020	Museums activities	4
8	91040	Botanical zoolog.gardens	3
9	79120	Tour operator activities	2
10	55200	Holiday accommodation	4

industries' complexity arising from clustering several industries (example Table 1) and stakeholders, dealing with a number of non-economic external variables (large range within tourism's primary resources—eg. natural resources, weather, seasonality etc.) and from the intangible nature of most products (leading to the necessity of more complex marketing insights), the scale of input data for analytics and decision making is several times larger. The tested dataset is already powering successful Web Report Servers (e. g. Finstat, Slovstat) that analyze mainly the profile and economic nature of enterprises. DBIS's aim is to provide a complex open solution for **DMOs** and their stakeholders with the possibility to create and combine their own data with DBIS's extracted data covering relations in tourism.

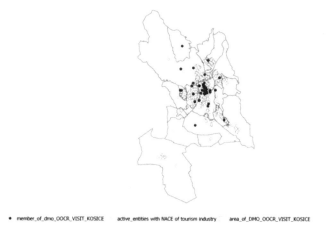

member_of_dmo_OOCR_VISIT_KOSICE   active_entities with NACE of tourism industry   area_of_DMO_OOCR_VISIT_KOSICE

Figure 3.   Comparison of DMO members and active tourism entities in destination Kosice (Source: Self-elaborated in QGis).

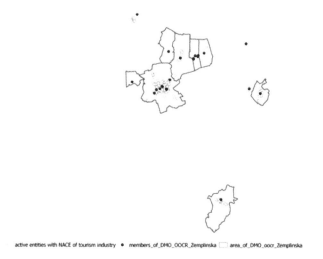

active entities with NACE of tourism industry   •   members_of_DMO_OOCR_Zemplinska   area_of_DMO_oocr_Zemplinska

Figure 4.   Comparison of DMO members and active tourism entities in destination Dolný Zemplín (Source: Self-elaborated in QGIS).

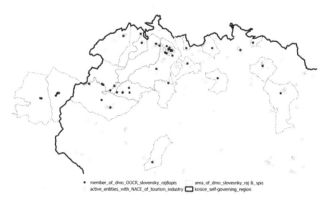

• member_of_dmo_OOCR_slovensky_raj&spis   area_of_dmo_slovesnky_raj &_spis
   active_entities_with_NACE_of_tourism_industry ☐ kosice_self-governing_region

Figure 5.   Comparison of DMO members and active tourism entities in destination Slovenský raj & Spiš (Source: Self-elaborated in QGIS).

4 CONCLUSIONS

The properties of the extracted data are high quality and work smoothly after merging with other open spatial data. The low share of Kosice region's DMOs' members' share on the total count of tourism industry entities is caused by the fact, that not all members use relevant main economic activity of the tourism industry. Thus entities that in fact conduct tourism related are not accounted by the chosen methodology. Elimination of these options may be achieved by generating secondary NACE codes based on the entities' business activities listings within the commercial register. All of the SQL scripts are available for back checking. Future diseminating of the results to end users will be carried out within DBIS's user interface via graphical dashboards and dynamic web maps.

ACKNOWLEDGMENT

This work was supported by the Slovak Research and Development Agency under the contract no. APVV-14-0797.

REFERENCES

Bulhalis, D. (2014). Working Definitions of Smartness and Smart Tourism Destination. Retrieved September, 2017, from http://t.co/xrLRpGipvu.

Delaney, J. and MacFeely, S. (2014). Extending Supply Side Statistics for the Tourism Sector: A New Approach Based on Linked-Administrative Data. Journal of the Statistical and Social Inquiry Society of Ireland, 43(1), pp.136–168. Retrieved September, 2017, from http://www.tara.tcd.ie/handle/2262/7277.

Dugas, J., Mesároš, P., Ferencz, V., Čarnický, Š., (2015). Business Communication Systems. 1st ed. Brusel: EuroScientia, 2015. 168 s. VEGA 1/0562/14. ISBN 978-90-822990-0-7.

Dupeyras, A., & MacCallum, N. (2013). Indicators for Measuring Competitiveness in Tourism. A Guidance Document (2013 No. 02). OECD Tourism Papers. Paris.

European Commission. (2008). RAMON – Reference and Management of Nomenclatures. Retrieved November 07, 2017, from http://ec.europa.eu/eurostat/ramon/nomenclatures.

European Commission. (2010). Communication from the Commission to the European Parliament, the Council, the European Economic and Social Committee and the Committee of the Regions—Europe, the world's No. 1 tourist destination—a new political framework for tourism in Europe. Retrieved September, 2017, from http://eur-lex.europa.eu/legal-content/EN/TXT/HTML/?uri=CELEX:52010DC0352.

European Commission (2016). The European tourism indicator system. Luxembourg: Publications Office.

European Union. (2014). Methodological manual for tourism statistics (3.1rd ed.). Luxembourg.

Fuchs, M., Abadzhiev, A., Svensson, B., Höpken, W. and Lexhagen, M. (2013). A knowledge destination framework for tourism sustainability: A business intelligence application from Sweden. Tourism: An International Interdisciplinary Journal, 61(2), pp.121–148. Retrieved September, 2017, from http://hrcak.srce.hr/index.php?show=toc&id_broj=8675.

Indicators of sustainable development for tourism destinations. (2004). Madrid, Spain: World Tourism Organization.

Ionescu, V. (2014). Are the tourism indicators able to quantify the regional disparities across the european union ? Retrieved September, 2017, from https://www.researchgate.net/publication/308468493.

Khouri, S., Al-Zabidi, D., Cehlár, M., & Alexandrová, G. (2009). The process of design of an efficient information system for a company. Journal of Engineering Annals of Faculty of Engineering Hunedoara, 7(3), 155–158.

Khouri, S., Cehlár, M., & Jurkasová, Z. (2011). Management Tool for Effective Decision—Business Intelligence. Journal on Law, Economy and Management, 1(1), 67–70. doi:ISSN: 2048-4186.

Kršák, B., Sidor, C., Štrba, Ľ., et al. (2015). Maximizing the potential of mining tourism through knowledge infrastructures. Acta Montanistica Slovaca, 20(4), 319–325, doi: 10.3390/ams20040319.

Kršák, B., Sidor, C., Štrba, Ľ. and Mitterpák, M. (2016). Usage of linked open data for the measurement of mining tourism POIs' impact on the competitiveness of a destination: Research notes part 1. Acta Montanistica Slovaca, 21(2), pp.162–169.

Kršák, B., Sidor, C. and Štrba, Ľ. (2016). Destination Business Information System—effective tool for tourism product developments. Book of Abstracts of the GEOTOUR 2016. Firenze: IBIMET-CNR, p.40.

McCole, D. and Joppe, M. (2014). The search for meaningful tourism indicators: the case of the International Upper Great Lakes Study. Journal of Policy Research in Tourism, Leisure and Events, 6(3), pp.248–263.

Muñoz, A.L., & Sánchez, S.G. (2013). Destinos turísticos inteligentes. Harvard Deusto business review (224), 58–67.

NT, B. (2017). Top Business Intelligence (BI) tools in the market. [online] Big Data Made Simple—One source. Many perspectives. Retrieved September, 2017, from http://bigdata-madesimple.com/top-business-intelligence-bi-tools-in-the-market.

Slovensko.digital. (2017). Otvorené dáta · Ekosystém.Slovensko.Digital. Retrieved October 12, 2017, from https://ekosystem.slovensko.digital/otvorene-data.

Štrba, Ľ., Kršák, B. and Sidor, C. (2016). Destinations business information systems for smart destinations: the case study of Kosice county. International Journal of Business and Management Studies, 5(1), pp.177–180.

United Nations (2008). Tourism satellite account: recommended methodological framework. Luxembourg.

Advances and Trends in Geodesy, Cartography and Geoinformatics – Molčíková et al. (Eds)
© *2018 Taylor & Francis Group, London, ISBN 978-1-138-58489-1*

Analysis of LiDAR data with low density in the context of its applicability for the cultural heritage documentation

T. Lieskovský, J. Faixová Chalachanová & L. Lessová
Faculty of Civil Engineering, Slovak University of Technology in Bratislava, Bratislava, Slovak Republic

M. Horňák
Slovak Association of Archaeologists, Banská Bystrica, Slovak Republic

ABSTRACT: The paper deals with assessment of utilization of airborne laser scanning data with low density for archaeological sites detection and analysis. Light Detection and Ranging (LiDAR) offers fast and quality method for data capture. The advantage of lower density preconditions for LiDAR data is that it results in decreasing processing time and data storage requirements. Within the area of interest, LiDAR data was obtained by airborne laser scanning realized by the National Forest Centre during the growing season with low density of points per m². Input point cloud data was filtered and then interpolated into the form of raster digital elevation model in order to compare it with reference etalon. The reference etalon was obtained by surveying using universal measuring station considering the morphology of the terrain. Measured points within the reference etalon was interpolated into the form of raster digital elevation model using ordinary kriging method. Both digital elevation models were compared and tested in terms of methodology in accordance with International Organization for Standardization (ISO) standard ISO 19157:2013 Geographic information—Data quality. Methods such as point-to-surface or surface-to-surface was used in the testing procedure. The paper is resulting into the recommendations for input conditions and criteria in the using of airborne laser scanning method with low density and within the densely-vegetated areas. Our study was realized in the field of archaeology in order to obtain information about the presence or absence of the archaeological site, as well as to obtain its suitable digital elevation model. Digital elevation model created from LiDAR data can then be used for the cultural heritage documentation and can be the basis for various spatial analysis, such as calculation of volume, relative height, visibility, etc.

1 INTRODUCTION

LiDAR technology provides fast and accurate alternative for mapping of an extensive areas. It is successively accepted as a basic technique for digital elevation model (DEM) generating. In the domain of cultural heritage documentation, the LiDAR technology is used to search for the remains of human settlements, whether hidden underground or scattered on the Earth's surface. According to (Xuelian et al. 2010), accuracy of ground filtering and terrain modeling is affected by several factors, and density of point cloud data is one of them. Influence of point cloud density to DEM is discussed in (Anderson et al. 2005). Requirement of higher point density for DEM with higher resolution was established in this study, as well as the influence of point density to linear interpolation methods for DEM creation (e.g. inverse distance) and kriging interpolation also. Several issues dealing with the influence of LiDAR dataset density on the efficiency of DEM generation (related to the amount of data and processing time) are dealt with (Liu et al. 2007). The result of LIDAR scanning is the point cloud with all measured points, including unwanted objects, so the decisive procedure in LiDAR data processing is filtering and classifying. According to (Chen et al. 2007) filtering includes methods based on interpolation, or morphological methods.

The aim of this study is to compare DEM obtained by LiDAR technology with DEM obtained by geodetic measurements using universal measuring station. The overall height accuracy of the both models was compared. The course of the profiles that capturing the significant elements of the fortification (valleys and trenches) was compared also. The volume of the mass for the selected area was calculated on selected profiles. As the result of this comparison, the recommendations for LiDAR data usage for the detection and documentation of archaeological structures (e.g. ravines and their elements) can be considered.

2 MATERIALS AND METHODS

2.1 *Materials*

The source point cloud data for comparison was obtained using LiDAR scanning performed by the National Forest Center (NFC). Area of the interest is situated in the territory of Neštich fort in the Bratislava region northwest of the village Svätý Jur. NLC data serve primarily for the forest management and protection, so their methodology and quality are adapted primarily for these purposes. The scanning was done on 29.08.2014, when the area was extensively covered with vegetation. Area of the interest is located in a deciduous forest. The ALS70-CM (City/Corridor Mapper) laser scanner was used to collect data. The scanning angle was 50°, the scanning frequency was 25,7 Hz. The reference height ranged from 168 to 599 m. The flight height was 1169–1600 m. The average point density is 2,13 points per square meter. The average distance between points is 0,68 m. Estimated accuracy in both, the transverse and longitudinal directions, is 0,15–0,19 m. Estimated height accuracy is 0,06–0,09 m. Input terrestrial data (approximately 3400 points) were provided by the Institute of Archeology of the Slovak Academy of Sciences in Nitra. Coordinates have been transformed from the SJTSK (Datum of Uniform Trigonometric Cadastral Network) and the Bpv (Baltic Vertical System after adjustment) elevation into the UTM 34 N (Universal Transverse Mercator zone 34 North) and ellipsoidal elevation. Terrestrial measured data was used as an etalon for comparison. It is a selective method where elements can be measured individually, for example based on the morphological background of the terrain (backbones, slugs), or on the characteristics that serve to the given purpose of mapping.

Basic processing of the point cloud data was realized in the software LAStools. Unclassified data passed through a processing process that begins with the identification of bareground points which are located directly on the terrain and non-ground points which are located off the terrain. The *lasground* tool was used, and a new file with classified terrain was created. Incorrectly determined points, that are too far off the terrain, were removed by the *lasheight* tool. The next step in the data preparation was the *lasclassify* tool, which classifies high vegetation (trees) and buildings. Data were analyzed in ArcGIS software and compared to the terrestrially measured data. The *las2dem* and *blast2dem* tools were used to create DEM in LAStools. The Fusion *GroundFilter* tool was used to filter the LiDAR point cloud data that lies on the assumed ground surface. Experiment has shown, that default coefficients for weight function produce reliable results only for high point density (>4 reflections/m²). At low point density, appropriate coefficients must be selected. The *GroundFilter* tool creates a set of points to create a surface that is suitable for calculating the height of vegetation. The output file can be used in *GridSurfaceCreate* tool or *TINSurfaceCreate* tool to create a ground surface model. To compare height profiles and cubatures, the terrestrially measured data were interpolated into the form of raster DEM at a resolution of 20 cm/pixel.

2.2 *Methods*

For data interpolation and DEM creation the method of ordinary kriging was used. According to (Vajsáblová et al. 2008), the ordinary kriging method, as well as the spline method and the nearest neighborhood method, are the best quality methods for the interpolation of archeological data. The resolution of 20 cm/pixel was chosen to get closer to the resolution of DEM processed by NLC. However, this value of resolution can be considered as a limit value, as the point density of geodetic survey was 0.19 points/m². Too high resolution at low input data density causes artifacts and distortions (i.e. bull eyes). Low resolution causes loss of detail, especially in areas with sharp changes in morphology of the relief (Vajsáblová et al. 2008).

The methodology of DEM testing is based on the standard ISO 19157:2013 Geographic information—Data quality, which sets out principles for the description of geographic data and its processing. By statistical calculations, we can determine the numerical deviations between the methods, their mean errors and the values of the maximum and minimum differences in heights. The point-to-point method was used to evaluate the interpolated DEM in raster or vector form (in our case LiDAR) against reference etalon points (in our case geodetic survey).

Another way of comparing the accuracy of LiDAR data was based on the profiles measurements, where was the points with high density terrestrially measured.

Processing of DEM and its comparison was realized in ArcGIS and AutoCAD Civil 3D software. Statistical calculations were performed in Microsoft Excel. The transformation of the Bpv heights into the ellipsoidal heights was realized by the transformation service provided by the Geodetic and Cartographic Institute (GCI). Testing for the presence of a systematic error by a Student's statistical test is intended to reveal a systematic shift in measurement in height. The mean difference between the height measured terrestrially and the height obtained by the LiDAR measurement was calculated, as well as the maximum and minimum height difference and the standard deviation also. From the measurements were excluded the outlying values outside the interval $(E - 3\sigma, E + 3\sigma)$, where E is the mean value of the difference in height between two methods and σ is the standard deviation of the height differences. To simplify the outlying values removal process, the limit value of 3σ was chosen, which is commonly used in practice for this purpose.

3 RESULTS

Analysis of the accuracy of LiDAR and terrestrial measured data took place in the following two steps:

1. Processing of all available measurements and statistical data calculation.
2. Exclusion of the outlying values of height differences with following suited data processing.

During data processing in the step 2), we excluded 3.3% of points that were detected as outlying values. The resulting average height difference between DEM obtained by LiDAR and DEM obtained by terrestrially surveying is 1,58 m below the terrestrial surface, and the standard deviation is 0,46 m (Table 1). Height differences are represented in the histogram (Figure 1). Student's t-test was used to test the presence of a systematic error at the selected significance level of 5% with the critical value t = 1,645. The null hypothesis was built as H_0: the mean value of height difference is equal to zero. Alternative hypothesis was built as H_1: the mean value of height difference is not equal to zero, so the systematic error can be assumed. The result of the Student's t-test was 144,823, so the zero hypothesis H_0 can be dismissed at the selected significance level of 5%, and the alternative hypothesis H_1 can be accepted, so the systematic error in the offset of measured data can be assumed.

Three control profiles were measured for the course of the height testing (Figure 2). Profile 1, located approximately in the middle of the area, is 70 meters long. Profile 2 is located at the

Table 1. Average height differences and its statistical characteristics for point-to-point method.

Processing step	Average height difference [m]	Standard deviation [m]	Maximal height difference [m]	Minimal height difference [m]
1	1,589	0,490	3,711	−2,555
2	1,579	0,461	3,054	−0,578

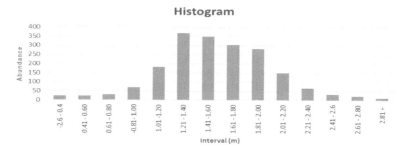

Figure 1. The histogram of height differences.

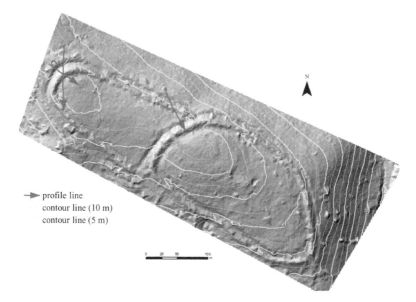

profile line
contour line (10 m)
contour line (5 m)

Figure 2. Graphical representation of the profiles of the Neštich site.

edge of the area and it is approximately 60 meters long. Profile 3, which crossing profile 2, is approximately 100 meters long. Measured profile points did not enter the previous calculation and can therefore be considered as an independent control. Each profile has been assigned with a LiDAR height value. The differences between the actual height and height measured in a given location from the LiDAR data were calculated. The comparison of calculated average height differences and their standard deviations for profiles 1, 2 and 3 is presented in Table 2.

Three control profiles were measured for the course of the height testing (Figure 2). Profile 1, located approximately in the middle of the area, is 70 meters long. Profile 2 is located at the edge of the area and it is approximately 60 meters long. Profile 3, which crossing profile 2, is approximately 100 meters long. Measured profile points did not enter the previous calculation and can therefore be considered as an independent control. Each profile has been assigned with a LiDAR height value. The differences between the actual height and height measured in a given location from the LiDAR data were calculated. The comparison of calculated average height differences and their standard deviations for profiles 1, 2 and 3 is presented in Table 2.

The results show that the greatest differences in the height between the terrestrial measured height and height obtained from the LiDAR measurement are located in places with low or almost not LiDAR data coverage areas, and in the area of trenches or valleys.

Surroundings of the measured site profiles, that pass through the valleys and the trenches are interesting in terms of ruggedness. Therefore, the area of ten meters and entire profile

Table 2. Average height differences and their statistical characteristics for profile method.

Profile number	Average height difference [m]	Standard deviation [m]	Maximal height difference [m]	Minimal height difference [m]	Number of measured points
1	1.693	0.3824	2.475	0.977	74
2	1.943	0.3698	3.159	1.048	86
3	2.014	0.3015	2.573	1.472	51

Table 3. Volumes calculated along the profiles.

Volume number	Volume value [m³]		Area [m²]	Ratio
	LiDAR	Terrestrial data		Terrestrial data/LiDAR
1	1878	2151	273	1.14
2	5096	5399	303	1.06
3	1256	1404	148	1.11

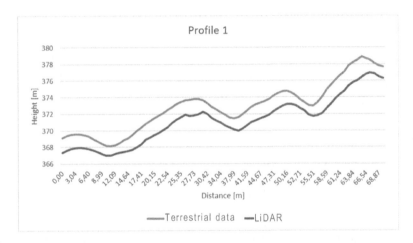

Figure 3. Representation of the profile heights with the profile lengths—profile 1.

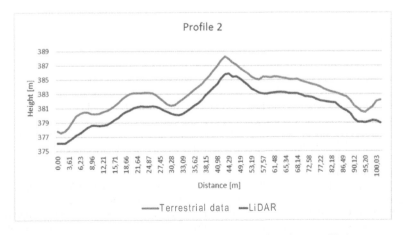

Figure 4. Representation of the profile heights with the profile lengths—profile 2.

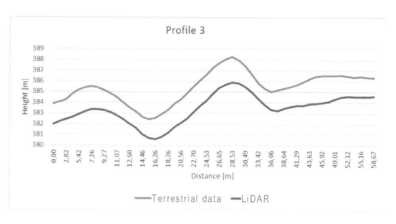

Figure 5. Representation of the profile heights with the profile lengths—profile 3.

length were used to calculate the volumes. The height from which the volume was calculated was determined from the average value of the profile. The average height difference between the height of the terrestrial and LiDAR measurements was added to the basic height. The volumes were calculated for a raster generated by Kriging interpolation at 50 cm step and for DEM obtained from LiDAR measurements. Values of the calculated volumes are presented in the Table 3. By comparing volumes, we have found relative accuracy between LiDAR and terrestrial measured data. From the volume ratio, the estimated accuracy of the LiDAR measurement is approximately 10% in relation to the terrestrially measured data. Figures 3–5 present the profiles heights with lengths of profiles calculated from points coordinates.

4 CONCLUSION

This study was intended to compare the quality of LiDAR data against the terrestrially measured data and verify the LiDAR data potential for the detection of objects falling under the domain of cultural heritage, as well as for the different types of spatial analyzes and calculations used in archaeological practice. Finding the suitability of large areas LiDAR scanning is a beneficial solution for large sites and densely wooded areas where terrestrial measurement is time-consuming and cost-effective. Therefore, this method of data collection could be appropriate for the preservation of cultural heritage. But the factors arising from the purpose of the scanning and following point density considerably influence the possibilities of using the data in question for archeology and morphometric analyzes.

REFERENCES

Anderson, E.S., Thompson, J.A., Crouse, D.A., Austin, R.E. 2005. Horizontal resolution and data density effects on remotely sensed LIDAR—based DEM. Geoderma 2005, 132, 406–415.

Chen, Q., Gong, P., Baldocchi, D., Xie, G. 2007. Filtering Airborne Laser Scanning Data with Morphological Methods. Photogrammetric Engineering & Remote Sensing, Number 2, American Society for Photogrammetry and Remote Sensing 2007, pp. 175–185.

Liu, X., Zhang, Z., Peterson, J., Chandra, S. 2007. The effects of LiDAR data density on DEM accuracy. In Proceedings of 2007 International Congress on Modeling and Simulation, Christchurch, New Zealand, 2007.

STN EN ISO 19157: 2013 Geographic information—Data quality. Slovak Office of Standards, Metrology and Testing, Bratislava 2013.

Vajsáblová, M., Fencík, R., Droppová, V. 2008. Metódy tvorby digitálnych modelov reliéfu. In Geometria a jej aplikácie 2008: Geometria krajiny a sídla, contributions in anthology, Bratislava 2008.

Xuelian, M., Nate, C., Kaiguang, Z. 2010. Ground Filtering Algorithms for Airborne LiDAR Data: A Review of Critical Issues. Remote Sensing 2010, 2(3), pp. 833–860. [cited 2017-06-20] <http://www.mdpi.com/2072–4292/2/3/833/htm>

Advances and Trends in Geodesy, Cartography and Geoinformatics – Molčíková et al. (Eds)
© *2018 Taylor & Francis Group, London, ISBN 978-1-138-58489-1*

Classification of digitized old maps

M. Talich & O. Böhm
Research Institute of Geodesy, Topography and Cartography, Zdiby, Czech Republic

L. Soukup
Institute of Information Theory and Automation of the CAS, Prague, Czech Republic

ABSTRACT: Because of their importance as historical sources, old maps are steadily becoming more interesting to researchers and public users. However, the users are no longer satisfied only by simple digitization and online publication. Users primarily require advanced web tools for more sophisticated work with old maps.

This paper is concerned with classification of digitized old maps in form of raster images. An automatic classification of digital maps is useful tools. This process allows to automatically detect areas with common characteristic, i.e. forests, water surfaces, buildings etc. Technically it is a problem of assigning the image's pixels to one of several classes defined in advance. If the map is georeferenced the classified image can be used to determine the surface areas of the classified regions, or otherwise evaluate their position.

Unfortunately quite substantial difficulties can be expected when attempting to apply these tools. The main cause of these difficulties is varied quality of digitized maps resulting from damage caused to the original maps by time or storage conditions and from varying scanning procedures. Even individual maps from the same map series can differ quite a lot.

The review of the main classification methods with special emphasis on the Bayesian methods of classification is given. An example of this classification and its use is also given. Web application of raster image classification is introduced as well. The web application can classify both individual images and raster data provided via Web Map Services (WMS) with respect to OGC standards (Open Geospatial Consortium). After gathering the data, classification is applied to distinguish separate regions in the image. User can choose between several classification methods and adjust pertinent parameters. Furthermore, several subsequent basic analytical tools are offered. The classification results and registration parameters can be saved for further use.

1 INTRODUCTION

Because of their importance as historical sources, old maps are steadily becoming more interesting to researchers and public users. However, users are no longer satisfied with only a simple digitization and online publications of them. Users primarily require advanced web tools for more sophisticated work with the digitized old maps.

One of the useful tools is automatic classification of old digitized maps. This process allows to automatically detect areas with common characteristic, i.e. forests, water surfaces, buildings etc. From technical point of view is it a problem of assigning the image's pixels to one of several classes defined in advance. If the map is georeferenced the classified image can be used to determine the surface areas of the classified regions, or otherwise evaluate their position.

Unfortunately, quite substantial difficulties can be expected when attempting to apply these tools. The main cause of these difficulties is varied quality of digitized old maps resulting from damage caused to the original maps by time or storage conditions and from varying scanning procedures. Even individual maps from a single map series can differ quite a lot.

The basic prerequisite for processing old maps in this way is to have them scanned and published online in standardized way, because only then the maps will be available easily enough to research, develop and try sufficiently robust and efficient solutions to presented problems. Classification of raster images of the old maps will be further discussed in chapter 2 with special emphasis on the Bayesian methods of classification.

2 CLASSIFICATION OF RASTER IMAGES OF THE DIGITIZED OLD MAPS

2.1 Problem formulation

Regions with some characteristic features have to be localized in a digital image of digitized old map. These features could be uniquely derived from the given attributes of pixels, e.g. color of a pixel. The whole image has to be decomposed into disjoint regions and each region has to be attributed by a unique class according to the prevalent characteristic features. The set of classes has to be given in advance. The decomposition task, i.e. classification, results in assignment a specified class to each pixel in the given digital image.

2.2 The required result

Result of classification has to be in form of a new image that consists of homogenous disjoint regions of different classes. The regions are distinguishable by class labels or colors that are explained in the associated legend.

2.3 Review of the main classification methods

Vast number of different classification methods have been designed during short history of development of computer image processing. Two main groups of classification methods can be recognized: deterministic and statistic. Other distinction between classification methods is based on practical circumstances of solution of the classification problem. When some characteristic features of the classes are available, the classification is called supervised. If no preliminary data about classes are known in advance, unsupervised classification (cluster analysis) has to be performed. Statistical supervised classification, see e.g. (Webb 2003), (Denison, Holmes, Mallick & Smith 2002), presents more powerful tool than the other kinds of classification.

Statistical characteristics of the all admissible classes have to be known at statistical supervised classification. The most common way of gaining characteristic features of classes is to determine training sets in the given image. Training set is a region in the image which well represents certain class. Searching for other regions with similar characteristic features is the task of supervised classification. Principle of statistical supervised classification is based on geometric notion of feature space. Feature space is an Euclidean space of points, whose coordinates are features that characterize each pixel in the image. Typical example of feature space is color space RGB. Each pixel of digital image displays as a point with coordinates that are the features, e.g. color components Red, Green, Blue. Points in feature space create clusters that represent particular classes. Some points of these clusters correspond to pixels that belong to a training set. These points can be labelled by identifier of a corresponding class. With the aid of the labelled points of training sets, other points in feature space have to be labelled to complete the classification. Hence the classification task can be formulated as a rule for labelling pixels displayed in feature space. This rule, called classifier, can be searched for by means of several manners. The most common classifiers are e.g. linear, nearest neighbor or Bayesian classifiers. Bayesian classification that is the most important member of the family of statistical supervised classification will be studied in the sequel.

2.4 Bayesian classification

2.4.1 Input data and assumptions

A digital image is given where training sets are determined. Certain number of classes is chosen to distinguish regions of different characteristics. Let $\mathcal{C}$ be the set of the all classes. Each training set is assigned to a certain class. Each class has to be represented by one training set at least. Furthermore a prior probability $P(C)$ has to be known for each class C in $\mathcal{C}$. The prior probabilities describe general preliminary information about presence of classes in the given image.

2.4.2 Solution of the problem

The classification problem is solved by Bayesian classifier in this contribution. The Bayesian classifier stems from Bayes formula (see e.g. (Webb 2003)). This formula enables to compute the probability that a pixel with feature vector F belongs to class C. It is conditional probability $P(C|F)$. We can estimate opposite conditional probability $P(F|C)$ for any feature vector F and class C with the aid of the training sets. Expression $P(F|C)$ stands for probability that a pixel of class C has feature vector F. Under these assumptions for known prior probabilities $P(C)$ the Bayes formula has form:

$$P(C|F) = \frac{P(F|C)P(C)}{\sum_{T \in \mathcal{C}} P(F|T)P(T)} \qquad (1)$$

The last step of the classification procedure comprises assignment of class C to pixel with feature vector F to maximize posterior probability $P(C|F)$.

Crucial problem resides in computation of probabilities $P(F|C)$, since it is sensitive to input data in training sets. Three variants of Bayesian classification will be presented to cover most cases of determining training sets.

2.4.3 Basic variant

The simplest way of computation probabilities $P(F|C)$ is based on relative frequency of pixels in the training set. Let us denote n_C the overall number of pixels in training set of class C and $n_{C,F}$ the number of pixels with feature vector F in the same training set. Then the probability $P(F|C)$ can be approximately estimated by

$$P(F|C) = \frac{n_{C,F}}{n_C} \qquad (2)$$

2.4.4 Extended variant

The extended variant is based on assumption, that clusters of the same class are normally distributed. Under this assumption each training set could be extended by adding other pixels with similar features as the original pixels selected by the actual training set in the chosen cluster.

Pixels, whose feature vectors are sufficiently close to the center of the cluster, could be treated as members of the actual class C. Such pixels can extend the actual training set to create a new, extended training set. The extended training set is more representative, but there is some risk that some of its pixels do not belong to the actual class C. If the risk is small (e.g. less than 0.05), it is possible to compute relative frequency (2) with greater values of n_C, $n_{C,F}$. Better estimation of probabilities $P(F|C)$ could be reached by this way. The problem is in definitions of riskiness and sufficient closeness to the center of cluster.

The distance of additional pixels from the center of cluster is measured by Mahalanobis distance. The limit distance below which the pixels are considered close has to be determined in accordance to the risk of appending wrong pixels.

2.4.5 Nearest neighbor variant

This variant is based on assumption of normality of clusters as in the previous variant. Indeed, membership of a pixel into a class is computed as a distance of the pixel from the center of the cluster. The pixel is assigned to the class, whose training set is the nearest to the pixel in question. The metrics for measuring the distance is derived from Mahalanobis distance. The distance between a pixel and a training set of class C is a posterior probability $P(C \mid E_h)$ which is given by Bayes formula in the consequent form.

$$P(P \mid E_h) = \frac{P(E_h \mid C)P(C)}{\sum_{T \in C} P(E_h \mid T)P(T)} \tag{3}$$

Symbol E_h depends on the risk of appending wrong pixels.

3 PRACTICAL ONLINE SOLUTION

Web application for practical solution of Bayesian classification was created. The application, named WACLASS, is available at www.vugtk.cz/ingeocalc/igc/classification/ and works as a client—server application.

The client part of the application supports all the user operations, namely design of classes, definition of training sets and so on. The actual classification runs on the server side of the application. This part of the application was programmed in Python language with the aid of web framework Django and image processing library PIL (Python Image Library). Client side of the application is based on standard web technologies such as HTML, Javascript, and SVG (Scalable Vector Graphics). It means that the application can be used on practically any computer that is connected to Internet with any web browser. More information about this web application with its features, data sources, possible variants of classification, analytical tools and application controls are in (Talich, Böhm & Soukup 2012).

4 PRACTICAL EXAMPLE

This chapter presents an example of use of automatic classification for estimating the surface area of former lake Štítarský in the first half of 19th century. This lake lay near present day Vinice village near Městec Králové town. The lake can be found on II. Military Survey maps, but it no longer exists today in its former size.

By comparing the old maps with contemporary maps it is possible to see that only small remnants of the original water body are left. The original size of the lake can be determined by using web application for raster image classification mentioned above and accessible on the www.vugtk.cz/ingeocalc. This application is part of a knowledge system for decision support based on geodata created in VÚGTK between years 2006 and 2011. The application can display (among others) the maps of II. Military Survey provided as WMS (OGC 2006) and use it as data source.

When the map is displayed it is necessary to select training areas—representative samples of the areas of interest. In this case, the area of interest is water surface and it is best represented by several rectangular areas inside the lake Štítarský (see Figure 1). Based on these training areas the application classifies the image, i.e. marks pixels with satisfying degree of similarity the pixels from training areas as water surface. Based on the characteristics of the processed image it might be necessary to try different classifying methods and/or adjust parameters of these methods.

When the result of classification is satisfying (see Figure 2), the surface areas of classified regions can be computed by using the "Statistics" utility from the "Tools" menu. This tool generates a table listing total surface areas of all classified regions in the image. In this case the area of water surfaces—the surface area of lake Štítarský in the first half of 19th century—is approximately 112 ha (hectare). For better demonstration a layer with contemporary map source can be displayed, for example the Basic Map of the Czech Republic 1:10 000 (as shown on Figure 3). The classified region then clearly designates the area where the lake used to be and changes to the area can be easily discerned.

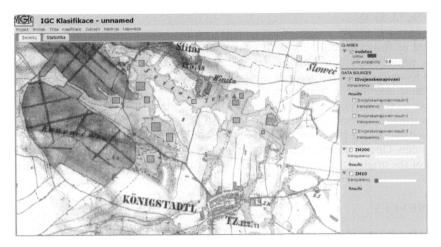

Figure 1.   Lake Štítarský, training areas selection.

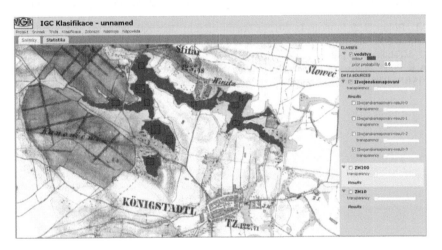

Figure 2.   Lake Štítarský, result of classification.

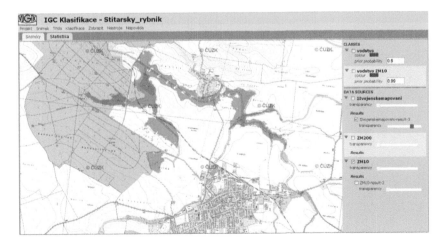

Figure 3.   Lake Štítarský, classification result displayed over contemporary map source.

5 CONCLUSION

The goal of this contribution was to point out the fact, that today users of the old maps are not content with bare online accessible maps any more. Nowadays an added value is required—online tools that allow to use the digitalized maps more efficiently, more easily and to get more information from them than from the original paper maps. One of such useful tools can be the automatic classification of old digitized maps.

This article discusses the methods for automatic classification of raster images of the old maps and presents a practical online tools and a practical example of use of classification of old maps. One of the practical results of such a classification can be, for example, the discovery of the original area of the water surface (lake) from the old map in a certain period.

REFERENCES

Denison D.G.T., Holmes C.C., Mallick B.K. & Smith A.F.M. 2002. *Bayesian Methods for Nonlinear Classification and Regression*. Willey series in probability and statistics. John Willey & Sons. ISBN: 978-0-471-49036-4.

OGC 2006. The OpenGIS® Web Map Service Interface Standard (WMS) http://www.opengeospatial.org/standards/wms.

Talich M., Böhm O. & Soukup L. 2012. Classification of digitized old maps and possibilities of its utilization. e-Perimetron, Volume 7, No. 3: 136–146, ISSN 1790-3769, http://www.e-perimetron.org/Vol_7_3/Talich_et_al.pdf.

Webb, A.R. 2003. *Statistical Pattern Recognition*. Second Edition, John Wiley & Sons, Ltd, Chichester, UK. ISBN: 9780470845134.

Author index